Learning Statistical Methods Using Statistical Software

Larry J Stephens

Steven From

Andrew Swift

and

Samantha Folken

Mathematics Department

University of Nebraska

Omaha, Nebraska 68182

Kendall Hunt
publishing company

This book has been printed from author prepared copy.

Background Cover and 2 right images © Shutterstock

Bottom Cover image © Andrew Swift

Kendall Hunt
p u b l i s h i n g   c o m p a n y

www.kendallhunt.com
Send all inquiries to:
4050 Westmark Drive
Dubuque, IA 52004-1840

Contents

Acknowledgements

Statistical software plays a large role in this book. We would like to thank the following people and the companies they are associated with for the use of their software in the book:

MINITAB – Linda Holderman, Coordinator of the author assistance program that Minitab sponsors. Portions of the input and output contained in this publication/book are printed with the permission of Minitab Inc. All material remains the exclusive property and copyright of Minitab Inc. The web address for Minitab is *www.minitab.com.*

SPSS – Jill Crist (formerly Rietema) SPSS, an IBM company. The following quote is from Wikipedia - **SPSS** is a computer program used for statistical analysis. Between 2009 and 2010 the premier software for SPSS was called PASW (Predictive Analytics Soft Ware) Statistics. The company announced July 28, 2009 that it was being acquired by IBM for US$1.2 billion. As of January 2010, it became "SPSS: An IBM Company". The web address for SPSS is www.spss.com. I would like to thank SPSS for permission to print from their software.

STATISTIX – Dr. Gerard Nimis, President, Analytical Software. I quote from the web site: "If you have data to analyze, but you're a researcher not a statistician , Statistix is designed for you. You'll be up and running in minutes without programming or using a manual. This easy to learn and simple to use software saves you valuable time and money. Statistix combines all the basic and advanced statistics and powerful data manipulation tools you need in a single, inexpensive package." The web address for STATISTIX is *www.statistix.com.* I would like to thank Dr. Nimis for permission to print from STATISTIX.

EXCEL – Excel has been around since 1985. It is available to almost all college students. It is widely used in this book.

We would like to thank Mary Dennison, Director of UNO Math Lab and Gary Meyer, Director of Information Technology College of Arts and Sciences at UNO for permission to use their Hi Tech graphic. I would also like to thank Jeffrey Huemoeller, Acquisitions editor at Kendall Hunt for all his help. And last but not least, my wife Lana Stephens for her patience and help.   Larry Stephens.

Preface

Hi Tech classrooms are used in a course which uses statistical software to learn statistical methods.

The student is encouraged to read the internet article entitled statistical methods and statistical computing found in Wikipedia.

If you have any comments concerning your experience in this course, please send an e-mail to me at Lstephens@unomaha.edu

Authors
Dr. Larry J. Stephens
Dr. Andrew Swift
Dr. Steven From
Samantha Folken

Chapter 1

What is Statistics? Statistics is the science of data. This involves collecting, organizing, summarizing, presenting, and analyzing data. What is Data? Data are facts (numerical or otherwise) that convey information from which conclusions can be drawn.

To most people, statistics means "numerical summaries", for example: the number of crimes reported in a city, or, the percentage of passes completed by a quarterback. These are examples of Descriptive Statistics, and while extremely useful, the science of statistics is much deeper.

One of the uses of Statistical Methods is to gather data on a (usually large) collection of objects, referred to as a Population. For example, the population of interest could be all the students at a university, or all the widgets produced by a particular factory. The only way to know everything about the population is to examine every member of it. This is called a Census, and it is the only way we can be absolutely sure about the exact nature of the population.

Unfortunately, a Census is not always feasible, usually because it would be too expensive or time consuming to conduct. If a census is not feasible, how can we still learn something about the population? An alternative to conducting a census is to select and examine a small collection of objects from the population in the hope that the data collected from those objects will tell us something about the population as a whole. This small collection of objects is called the Sample, and the act of choosing them is called Sampling.

Without a census, we are unable to describe the population with certainty. However if we have sample data taken from the population, we will (with the help of some statistical methods) be able to infer things about the population. This is known as Statistical Inference.

How is a sample chosen from the population? There are many techniques for choosing a sample, however, many of them have flaws that lead to inaccuracies in the inference. To avoid this, Random Sampling should be used. That is, members of the sample are chosen at random from the population.

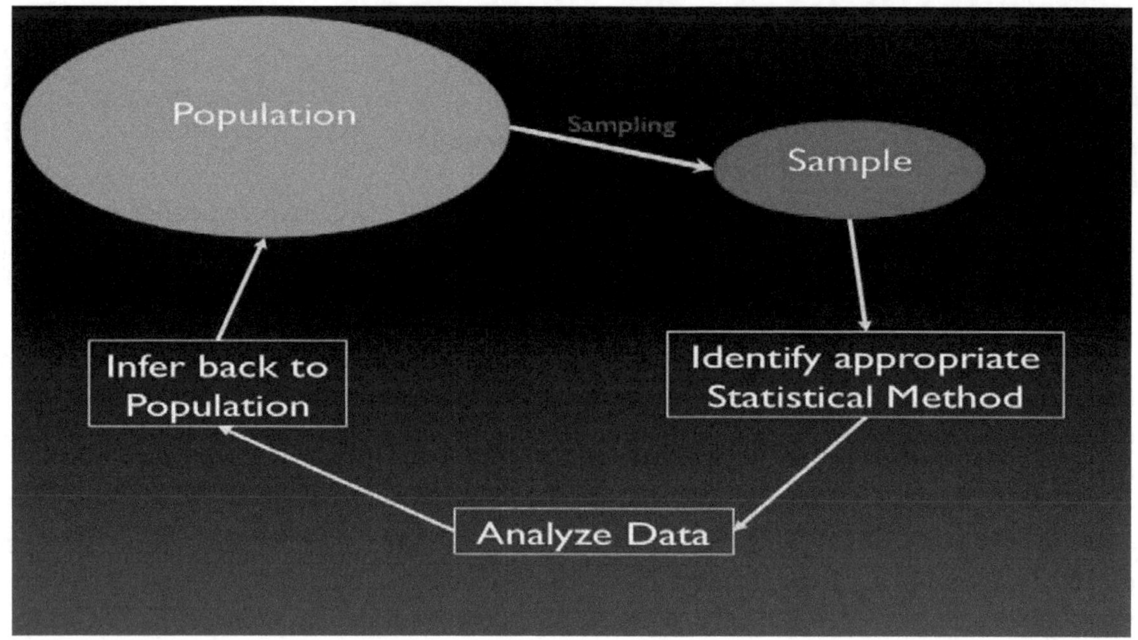

     This book introduces a new and innovative way of teaching Statistical Methods. The topics in Statistical Methods are taught by the use of Statistical Software. The Statistical Software that is used is EXCEL, SPSS, MINITAB, and STATISTIX. The student will learn Statistical Software alongside Statistical Methods rather than as an afterthought.

     In this Chapter, we introduce graphical methods for describing data. In Chapters 2 and 3 we discuss numerical summaries of the data. In Chapters 4 and 5 we discuss probability, the mathematics of randomness and chance, the relevance of which is necessitated by our use of random sampling. Finally, in Chapters 6 and 7, we discuss some techniques for performing statistical inference.

     To begin the discussion of statistical methods, let us introduce some data. First, a statistical data file is introduced. The **data file** that is used is called spring 13.xls and is as follows.

A	B	C	D	E	F	G	H	I	J
Gender	class	age	gpa	ht	wt	Internet	phone	exercise	party
F	f	19	3	67	160	9	4	30	other
F	f	18	3.2	67	150	10	7	240	other
F	f	18	4	62	110	10	5	120	other
F	f	18	3.9	68	135	10	25	650	republican
F	f	18	3.8	67	135	10	15	150	republican
F	f	19	3.7	64	115	7	14	120	republican

F	f	18	3.5	74	145	5	2	240	other
F	f	18	3	68	120	7	8	120	republican
F	f	18	3.5	65	133	20	20	700	republican
F	f	18	4	64	148	40	1	60	other
F	f	19	3.4	64	115	20	15	300	republican
F	f	19	3	67	160	9	4	30	other
F	f	18	3.2	67	150	10	7	240	other
F	f	18	4	62	110	10	5	120	other
F	f	18	3.9	68	135	10	25	650	republican
F	f	18	3.8	67	135	10	15	150	republican
F	f	19	3.7	64	115	7	14	120	republican
F	f	18	3.5	74	145	5	2	240	other
F	f	18	3	68	120	7	8	120	republican
F	g	25	3.6	68	250	24	50	90	republican
F	g	48	3	63	200	10	7	270	democrat
F	g	44	4	67	168	4	2	120	other
F	g	50	3.8	65	159	4	3	90	democrat
F	g	27	3	67	160	20	10	30	other
F	g	25	3.6	68	250	24	50	90	republican
F	g	48	3	63	200	10	7	270	democrat
F	j	35	3.9	68	154	3	3	60	other
F	j	20	2.9	64	142	6	1	420	democrat
F	j	20	3.1	67	150	12	3	360	other
F	j	20	3	67	170	20	10	560	other
f	j	24	3.5	65	163	5	1	240	democrat
f	j	21	3.8	66	165	4	2	300	republican
f	j	21	3.6	65	125	4	3	0	republican
f	j	25	3.2	67	135	10	20	600	democrat
f	j	35	3.9	68	154	3	3	60	other
f	j	20	2.9	64	142	6	1	420	democrat
f	j	20	3.1	67	150	12	3	360	other
f	j	20	3	67	170	20	10	560	other
f	j	24	3.5	65	163	5	1	240	democrat
f	j	21	3.8	66	165	4	2	300	republican
f	j	21	3.6	65	125	4	3	0	republican
f	s	20	3.9	62	130	25	10	180	republican
f	s	21	2.8	63	190	39	1	120	republican
f	s	20	3.8	61	96	6	3	60	other
f	s	31	2.5	62	125	10	1	0	democrat
f	s	21	3.8	61	100	8	2	60	republican
f	s	20	2.6	67	160	28	15	0	other
f	s	20	3.7	63	145	8	1	180	other

f	s	20	3	63	150	18	-	2	120	other
f	s	20	3.5	70	125	20		20	210	other
f	s	43	3.7	67	280	32		21	185	democrat
f	s	20	2.9	63	140	20		2	500	other
f	s	24	2.7	66	150	16		3	180	other
f	s	20	3.9	62	130	25		10	180	republican
f	s	21	2.8	63	190	39		1	120	republican
f	sr	22	3	64	129	9		4	240	other
f	sr	21	3.9	66	160	10		10	120	democrat
f	sr	26	2.6	65	150	25		2	60	democrat
f	sr	23	3.4	67	135	16		8	360	republican
f	sr	27	2.9	65	135	2		2	60	democrat
f	sr	24	3.2	62	116	10		8	200	other
f	sr	20	2.9	69	139	10		15	240	democrat
f	sr	24	4	64	140	10		7	380	republican
f	sr	41	3.3	64	240	32		24	140	other
f	sr	21	3.3	61	120	30		30	60	republican
f	sr	21	3.8	67	160	25		10	100	other
f	sr	27	3.6	69	160	3		3	150	republican
f	sr	22	3	64	129	9		4	240	other
f	sr	21	3.9	66	160	10		10	120	democrat
f	sr	26	2.6	65	150	25		2	60	democrat
f	sr	23	3.4	67	135	16		8	360	republican
f	sr	27	2.9	65	135	2		2	60	democrat
m	j	22	3.5	75	190	20		1	1200	other
m	j	20	3.1	64	120	10		1	60	republican
m	j	22	3.4	74	185	10		5	250	other
m	j	22	3.2	71	109	38		4	50	other
m	j	20	3.2	72	150	6		14	240	republican
m	j	35	3	70	180	3		4	200	republican
m	j	20	3.3	72	270	5		2	360	other
m	j	28	2.9	69	172	5		4	500	other
m	j	24	3.3	66	140	5		2	300	democrat
m	s	20	4	72	180	17		2	450	democrat
m	s	20	3	70	165	10		2	700	republican
m	s	20	3	73	180	20		20	360	republican
m	s	23	3.1	75	175	21		2	150	republican
m	s	19	3	72	130	14		30	600	republican
m	sr	21	3.4	72	170	25		1	120	other
m	sr	23	3.2	72	180	16		8	240	other
m	sr	25	3.3	75	230	20		5	240	democrat
m	sr	26	3.4	75	200	4		1	45	other

m	sr	27	3.9	67	160	10	1	180	democrat
m	sr	30	3.6	72	180	6	1	300	other
m	sr	32	2.9	76	300	8	1	120	republican
m	sr	22	3.9	66	220	4	1	320	other
m	sr	23	2.8	71	182	14	2	360	republican
m	sr	23	3.2	68	190	10	1	1500	democrat
m	sr	27	3.1	74	300	28	15	120	other

This ***data file*** contains 10 ***variables*** and 97 records. Three of the variables are ***qualitative*** or result in nonnumeric data and 7 are ***quantitative*** or result in numeric data. The variables age, gpa, ht, wt, internet, phone, and exercise are quantitative and the variables gender, class, and party are qualitative. Age is a whole number, gpa is number less than 4.00, ht is a whole number, wt is a whole number, internet is a whole number indicating the hours per week spent on the internet, phone is a whole number indicating hours per week spent in phone conversation and exercise is a whole number indicating the minutes per week spent at the university gym. Gender takes on the categorical values m and f, class takes on the values f, s, j, sr, and g and party takes on the categorical values democrat, republican or other. We shall use this data file and a software package to analyze it. The data were collected from students in a statistical methods course.

Graphical output teaches us a lot about statistics. Let's consider the variable age in this data file. What shape does the distribution of ages have in this file? Suppose we use the software MINITAB to tell us about the ages in this file. Suppose the file spring13.xls is opened in MINITAB, we use the pull down **Graph → Dot plot** to form the following graphic for ages. This ***dot plot is formed by placing a dot above each age in the file.*** It tells us that most of the students are in their late teens and twenties. A few are in their thirties and forties. This distribution has a long tail to the right or is said to be ***skewed to the right***.

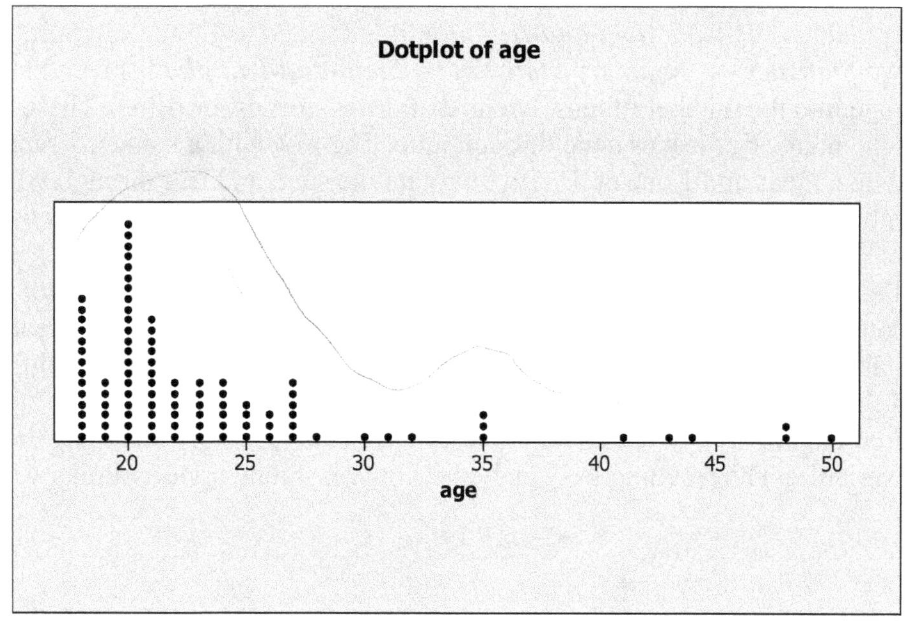

Statistix 9.0   2/8/2013, 2:32:46
PM

Stem and Leaf Plot of age

```
  Leaf Digit Unit = 1                    Minimum   18.000
  1  8   represents 18.                  Median    21.000
                                         Maximum   50.000

            Stem   Leaves
     20       1    8888888888888888999999
    (33)      2    000000000000000000001111111111111
     44       2    222222333333
     32       2    4444445555
     22       2    666777777
     13       2    8
     12       3    01
     10       3    2
      9       3    555
      6       3
      6       3
      6       4    1
      5       4    3
      4       4    4
      3       4
      3       4    88

      1       5    0
```

97 cases included   0 missing cases

    This graphic is called a ***stem-and-leaf plot***. It is formed using the software STATISTIX. The pull down ***Statistics → Summary Statistics → Stem-and-leaf plot*** is given to form it. It breaks each age into the number of tens, twenties, thirties, forties, or fifties. These are the stems. It also gives the number of leaves none through nine. The row with a 4 and a 1 represents an individual of age 4 tens and 1 one or 41. If you rotate the stem and leaf through 90 degrees you see the resemblance between the stem and leaf and the dot plot.

    To illustrate graphical techniques applied to qualitative variables, consider the use of EXCEL to construct a ***pie chart***. Open the data file as an EXCEL worksheet. Create a frequency distribution table by listing each class in column L of the EXCEL worksheet and frequency in column K of the worksheet. In the frequency for freshman, input the function =COUNTIF(column b, "f"). Use this function for each class category changing the "f" to the appropriate variable. The resulting frequency distribution should appear similar to the following.

class	Frequency
freshmen	19
sophomore	19
junior	24
senior	28
graduate	7

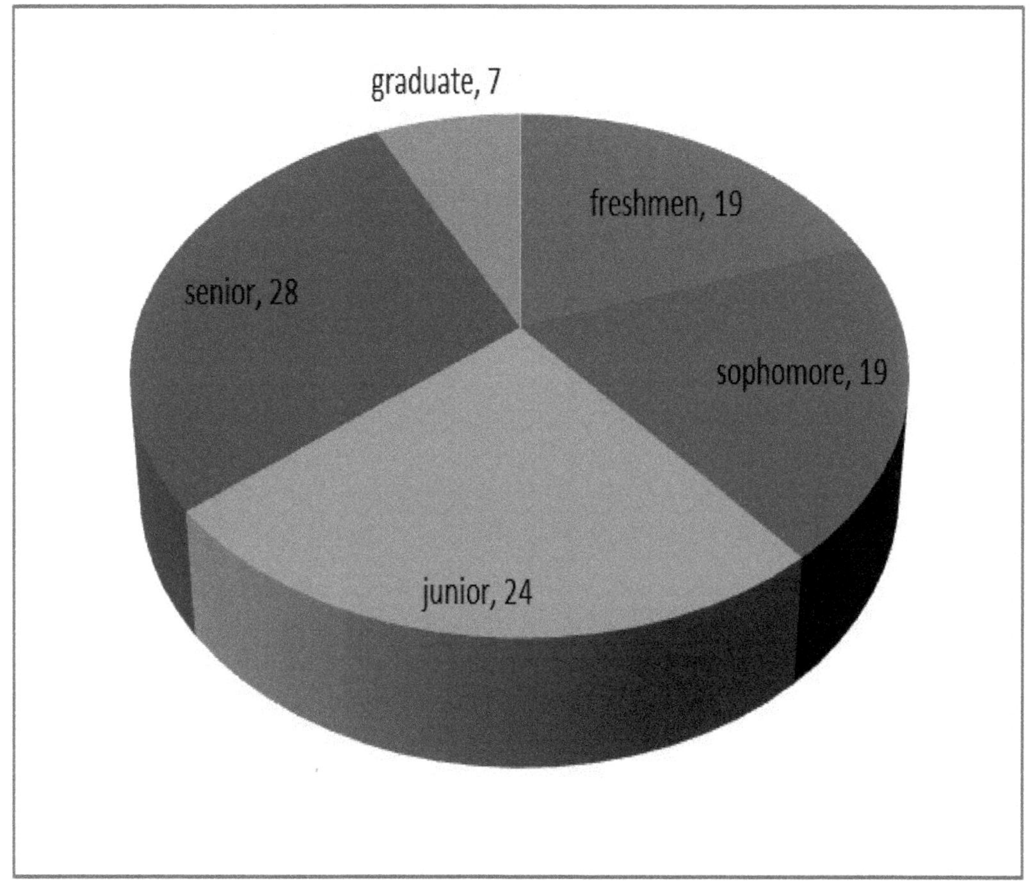

A ***percent distribution*** replaces the frequencies with percents. This is shown in the following pie chart.

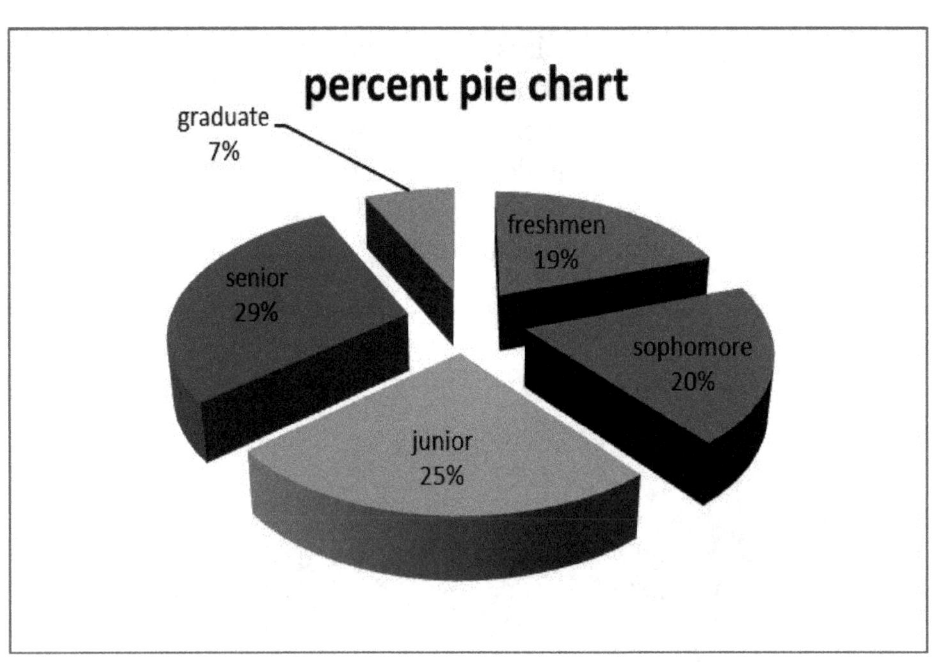

percent pie chart
graduate
7%
senior
29%
freshmen
19%
sophomore
20%
junior
25%

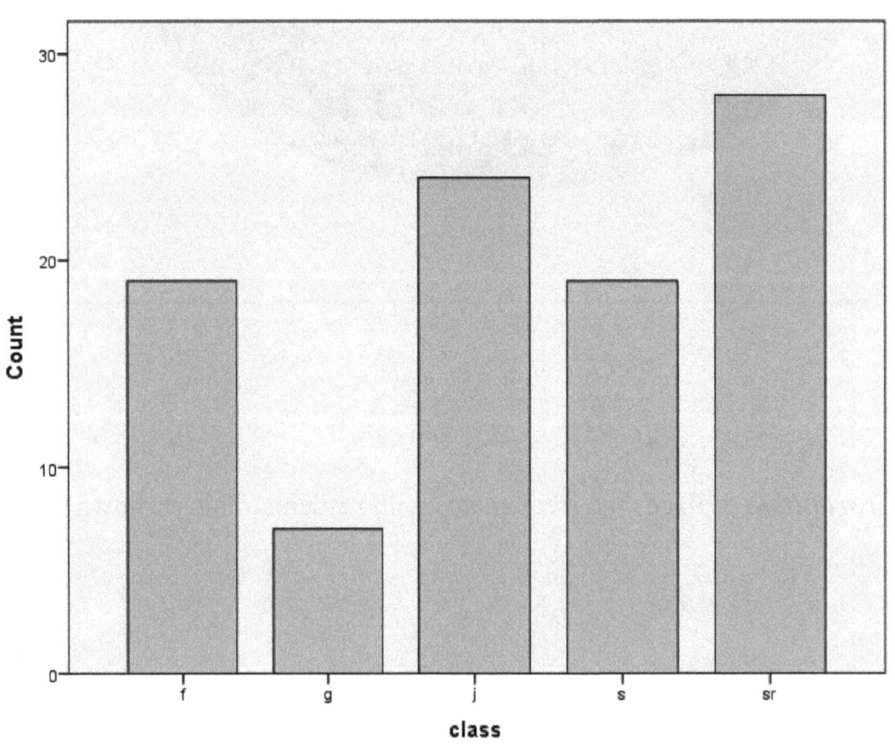

30
20
10
0
Count
f
g
j
s
sr
class

The SPSS *bar chart consists of bars whose heights are the frequency of the class and whose base is the class.* The data file is read into SPSS and the pull down **graphs → Legacy dialog → bar** is given to produce the above output.

This introductory chapter has introduced the student to statistical methods. The student has been introduced to graphical techniques by the use of four statistical software packages: EXCEL, STATISTIX, MINITAB, and SPSS. Of course, you may use another software package.

The student has been introduced to various statistical definitions and terms: These include:data file, variable, qualitative variable, quantitative variable, dot plot, skewed distribution to the right, stem-and-leaf plot, pie chart, frequency, frequency distribution, percent distribution, bar chart.

Problems

Refer to the data file Spring13.xls to do the following problems.

1.   Construct a dot plot for the variable time spent on the internet and print out the graphic.

   What value has the greatest frequency?

   What shape does the distribution have?

2.   Construct a stem-and-leaf plot for the weights and print out the plot.

   What value occurs the most often?

   How many distinct numbers are represented in the plot?

3.   Build a frequency distribution for the variable party.

   Construct a pie chart that includes each member of the party and the frequency of the party.

   Print out the pie chart.

4.   Construct a pie chart that includes each member of the party and the percent that the member makes up of that piece of the pie. What is the angle for each slice of the pie?

5.   Construct a bar chart for the frequency of each member of the party. Print out the bar chart.

   How tall is each bar?

Chapter 2

There are measures of data sets that are single numbers that in some sense represents the whole set of data. There are many such measures. We will define and discuss three of these measures. The first measure is the mean. There are several symbols for representing the mean. We shall represent the mean by the letter M. It is also represented by X with a bar on top (Xbar). M is ***the sum of the data divided by the amount of data, represented by n.***

$$M = \sum X / n$$

If EXCEL is used to sum the weights, we obtain $\sum X$ = 154.58. If we then divide by the number of weights in the file, n = 97, we obtain $M = \sum X / n$ = 159.3608. EXCEL also has a **built-in function** for the mean. It is called by typing the expression =AVERAGE(F2:F98), which also gives 159.3608.

The ***median is the middle number of a data set***. If the amount of data is odd, the median is the middle number. If the amount of data is even, the median is the average of the two numbers closest to the middle. An important operation is ***sorting the data*** from smallest to largest. If the weights are sorted and printed out, the result is 150 pounds. The middle of 97 numbers is the 49th number, or 150 pounds. The median is also given by the built-in function = Median(F2:F98) or 150 pounds.

The ***Mode is the most frequently occurring data value in the data file***. In this file there are 3 modal values. They are 135, 150, and 160. They each occur nine times. The weight values in this file are ***tri-modal***.

Other descriptive statistics are the Geometric mean, the Harmonic mean and Percentiles.

Each software package gives a comprehensive set of Descriptive Statistics. The EXCEL pull down menu **Data → Data Analysis → Descriptive Statistics** gives the following output.

	Weight
Mean	159.3608
Standard Error	4.129045
Median	150
Mode	160
Standard Deviation	40.66637
Sample Variance	1653.754
Kurtosis	2.819295
Skewness	1.540673
Range	204
Minimum	96
Maximum	300
Sum	15458
Count	97

The data is entered into the worksheet of the package STATISTIX and the pull down **Statistics → Summary Statistics → Descriptive Statistics** gives the following STATISTIX output.

Statistix 9.0   2/10/2013, 3:00:22

Descriptive Statistics

	wt
N	97
Sum	15458
Mean	159.36
SD	40.666
Variance	1653.8
SE Mean	4.1290
Minimum	96.000
1st Quarti	135.00
Median	150.00
3rd Quarti	173.50
Maximum	300.00
MAD	20.000
Skew	1.5167
Kurtosis	2.6148

The data is read into the MINITAB worksheet. The pull down **Stat → Basic Statistics → Display Descriptive Statistics** gives the following output.

Descriptive Statistics: wt

```
              Total
Variable      Count    N   N*   CumN   Percent   CumPct     Mean   SE Mean   TrMean   StDev
wt               97   97    0     97       100      100   159.36      4.13   155.55   40.67


Variable   Variance   CoefVar       Sum   Sum of Squares   Minimum       Q1   Median
wt          1653.75     25.52   15458.00       2622160.00     96.00   135.00   150.00

                                                            N for
Variable         Q3   Maximum    Range     IQR            Mode     Mode   Skewness
wt           173.50    300.00   204.00   38.50   135, 150, 160        9       1.54


Variable   Kurtosis       MSSD
wt             2.82    1225.67
```

If the data is entered into the IBM SPSS Statistics Data Editor, use the pull down **Analyze → Descriptive → Descriptive Statistics** gives the following SPSS output.

Descriptive Statistics

	N	Minimum	Maximum	Sum	Mean	Std. Deviation	Variance
Wt	97	96	300	15458	159.36	40.666	1653.754
Valid N (listwise)	97						

Data may be expressed as ***standardized values*** or ***Z-values***. A data value, X, is expressed as a standardized value as follows.

$$Z = (X - M) / S$$

Where Z is the standardized value, X is the value being standardized, M is the mean for the group of values being standardized and S is the standard deviation for the data.

Express the male heights as standardized values. The male heights are entered into column C1 of the MINITAB worksheet.

<u>male heights</u>
75
64
74
71
72
70
72
69
66
72
70
73
75
72
72
72
75
75
67
72
76
66
71
68
74

The mean and Standard deviation of the male weights are found by using Descriptive statistics.

Descriptive Statistics: male heights

```
Variable          N  N*     Mean  SE Mean   StDev  Minimum        Q1  Median        Q3
male heights     25   0   71.320    0.640   3.198   64.000    69.500  72.000    74.000

Variable       Maximum
male heights    76.000
```

We see that M = 71.320 and S = 3.198. We wish to convert the heights to Z-values. Use the MINITAB calculator to convert the male heights to Z-values. Give the pull down Calc → Calculator and fill in the dialogue box as follows:

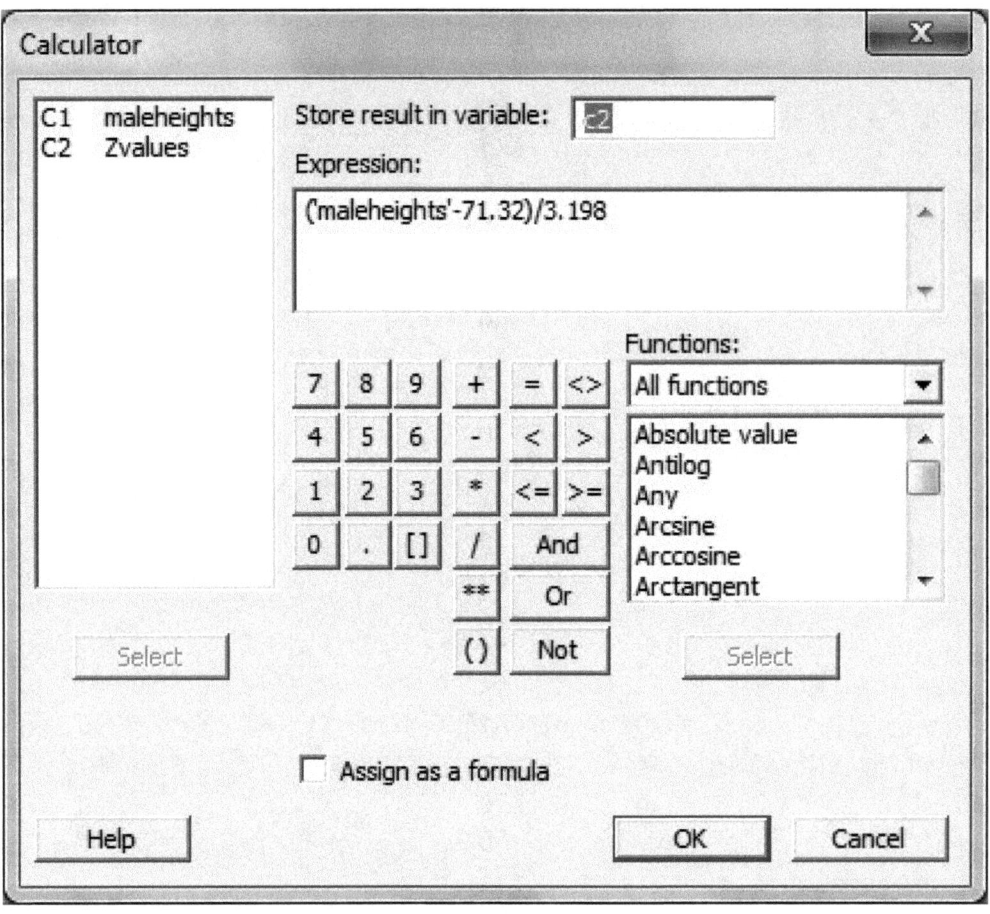

When OK is clicked the following is obtained.

maleheights	Zvalues
75	1.15072
64	-2.28893
74	0.83802
71	-0.10006
72	0.21263
70	-0.41276
72	0.21263
69	-0.72545
66	-1.66354
72	0.21263
70	-0.41276

73	0.52533
75	1.15072
72	0.21263
72	0.21263
72	0.21263
75	1.15072
75	1.15072
67	-1.35084
72	0.21263
76	1.46341
66	-1.66354
71	-0.10006
68	-1.03815
74	0.83802

The male who is 64 inches tall is 2.28893 standard deviations below the average of the group of males and the male who is 75 inches is 1.150 inches above the average of the group. A negative z-value results when the number is below the mean and a positive z-value occurs when the number is above the mean.

The descriptive statistics for the Zvalues is found using MINITAB to be as follows:

```
Variable  N  N*  Mean  SE Mean  StDev  Minimum    Q1 Median   Q3
Zvalues  25  0  0.000  0.200    1.000  -2.289    -0.569 0.213 0.838

Variable  Maximum
Zvalues    1.463
```

Note that the mean of the Z-values is 0 and the standard deviation is 1. This is always the case.

The following terms have been introduced in chapter 2: **Mean, Median, Mode**, the EXCEL pull down **Data → Data Analysis → Descriptive Statistics**, the STATISTIX pull down **Statistics → Summary Statistics → Descriptive Statistics,** the MINITAB pull down **Stat → Basic Statistics → Display Descriptive Statistics,** the SPSS pull down **Analyze → Descriptive → Descriptive Statistics.** Also, standardized or **Z-values** have been introduced.

Problems

For the Internet times in the file Spring13.xls, find the following.

1. M, the mean internet time.

2. The median internet time.

3. Find the modal internet time or times.

Print your output for 4 through 8.

4. Find the internet descriptive statistics using EXCEL.

5. Find the internet descriptive statistics using STATISTIX.

6. Find the internet descriptive statistics using MINITAB.

7. Find the internet descriptive statistics using SPSS.

8. Find the Z values for the graduate students in the file. Give the mean and standard deviation for the Z values.

Chapter 3

In the file spring13.xls, the variable exercise is much more variable than height. One measure of variability is the Range. The range is defined to be

$$\textbf{\textit{Range = Max – Min.}}$$

Suppose we read the file spring13.xls into EXCEL. We obtain the following the following records.

A	B	C	D	E	F	G	H	I	J
gender	class	age	gpa	ht	wt	internet	phone	exercise	party
f	f	19	3	67	160	9	4	30	other
f	f	18	3.2	67	150	10	7	240	other
f	f	18	4	62	110	10	5	120	other
f	f	18	3.9	68	135	10	25	650	republican
f	f	18	3.8	67	135	10	15	150	republican
f	f	19	3.7	64	115	7	14	120	republican
f	f	18	3.5	74	145	5	2	240	other
f	f	18	3	68	120	7	8	120	republican
f	f	18	3.5	65	133	20	20	700	republican
f	f	18	4	64	148	40	1	60	other
f	f	19	3.4	64	115	20	15	300	republican
f	f	19	3	67	160	9	4	30	other
f	f	18	3.2	67	150	10	7	240	other
f	f	18	4	62	110	10	5	120	other
f	f	18	3.9	68	135	10	25	650	republican
f	f	18	3.8	67	135	10	15	150	republican
f	f	19	3.7	64	115	7	14	120	republican
f	f	18	3.5	74	145	5	2	240	other
f	f	18	3	68	120	7	8	120	republican
f	g	25	3.6	68	250	24	50	90	republican
f	g	48	3	63	200	10	7	270	democrat
f	g	44	4	67	168	4	2	120	other
f	g	50	3.8	65	159	4	3	90	democrat
f	g	27	3	67	160	20	10	30	other
f	g	25	3.6	68	250	24	50	90	republican
f	g	48	3	63	200	10	7	270	democrat
f	j	35	3.9	68	154	3	3	60	other
f	j	20	2.9	64	142	6	1	420	democrat
f	j	20	3.1	67	150	12	3	360	other
f	j	20	3	67	170	20	10	560	other
f	j	24	3.5	65	163	5	1	240	democrat

f	j	21	3.8	66	165	4	2	300	republican
f	j	21	3.6	65	125	4	3	0	republican
f	j	25	3.2	67	135	10	20	600	democrat
f	j	35	3.9	68	154	3	3	60	other
f	j	20	2.9	64	142	6	1	420	democrat
f	j	20	3.1	67	150	12	3	360	other
f	j	20	3	67	170	20	10	560	other
f	j	24	3.5	65	163	5	1	240	democrat
f	j	21	3.8	66	165	4	2	300	republican
f	j	21	3.6	65	125	4	3	0	republican
f	s	20	3.9	62	130	25	10	180	republican
f	s	21	2.8	63	190	39	1	120	republican
f	s	20	3.8	61	96	6	3	60	other
f	s	31	2.5	62	125	10	1	0	democrat
f	s	21	3.8	61	100	8	2	60	republican
f	s	20	2.6	67	160	28	15	0	other
f	s	20	3.7	63	145	8	1	180	other
f	s	20	3	63	150	18	2	120	other
f	s	20	3.5	70	125	20	20	210	other
f	s	43	3.7	67	280	32	21	185	democrat
f	s	20	2.9	63	140	20	2	500	other
f	s	24	2.7	66	150	16	3	180	other
f	s	20	3.9	62	130	25	10	180	republican
f	s	21	2.8	63	190	39	1	120	republican
f	sr	22	3	64	129	9	4	240	other
f	sr	21	3.9	66	160	10	10	120	democrat
f	sr	26	2.6	65	150	25	2	60	democrat
f	sr	23	3.4	67	135	16	8	360	republican
f	sr	27	2.9	65	135	2	2	60	democrat
f	sr	24	3.2	62	116	10	8	200	other
f	sr	20	2.9	69	139	10	15	240	democrat
f	sr	24	4	64	140	10	7	380	republican
f	sr	41	3.3	64	240	32	24	140	other
f	sr	21	3.3	61	120	30	30	60	republican
f	sr	21	3.75	67	160	25	10	100	other
f	sr	27	3.6	69	160	3	3	150	republican
f	sr	22	3	64	129	9	4	240	other
f	sr	21	3.9	66	160	10	10	120	democrat
f	sr	26	2.6	65	150	25	2	60	democrat
f	sr	23	3.4	67	135	16	8	360	republican
f	sr	27	2.9	65	135	2	2	60	democrat
m	j	22	3.5	75	190	20	1	1200	other

m	j	20	3.1	64	120	10	1	60	republican
m	j	22	3.4	74	185	10	5	250	other
m	j	22	3.2	71	109	38	4	50	other
m	j	20	3.2	72	150	6	14	240	republican
m	j	35	3	70	180	3	4	200	republican
m	j	20	3.3	72	270	5	2	360	other
m	j	28	2.9	69	172	5	4	500	other
m	j	24	3.3	66	140	5	2	300	democrat
m	s	20	4	72	180	17	2	450	democrat
m	s	20	3	70	165	10	2	700	republican
m	s	20	3	73	180	20	20	360	republican
m	s	23	3.1	75	175	21	2	150	republican
m	s	19	3	72	130	14	30	600	republican
m	sr	21	3.4	72	170	25	1	120	other
m	sr	23	3.2	72	180	16	8	240	other
m	sr	25	3.3	75	230	20	5	240	democrat
m	sr	26	3.4	75	200	4	1	45	other
m	sr	27	3.85	67	160	10	1	180	democrat
m	sr	30	3.6	72	180	6	1	300	other
m	sr	32	2.9	76	300	8	1	120	republican
m	sr	22	3.85	66	220	4	1	320	other
m	sr	23	2.75	71	182	14	2	360	republican
m	sr	23	3.2	68	190	10	1	1500	democrat
m	sr	27	3.1	74	300	28	15	120	other

The range for the variable exercise is = Max(I2:I98) – Min(I2:I98) = 1500 – 0 = 1500 minutes.

The range for the variable height is = Max(E2:E98) – Min(E2:E98) = 76 – 61 = 15 inches. These commands are typed in any open cell in the EXCEL worksheet.

Suppose we compare the range in heights for males and females.

The range for males is = Max(E74:E98) – Min(E74:E98 = 76 – 64 = 12 inches and the range for females is = Max(E2:E73) – Min(E2:E73) = 74 – 61 = 13 inches

The range is affected by extreme values. This is a short-coming of the range as a measure of variability or dispersion.

The most widely used measure of dispersion or variability is variance or standard deviation. Before defining the variance the deviation of a variable is defined. ***The deviation of a variable from the mean is defined as***

$$(X - M).$$

For the freshmen in the file Spring13.xls, the deviations of the weights from the mean of the freshmen weights are as follows. The mean of the freshmen weights is 133.4737 pounds. The deviations from the mean are

$$\underline{X - 133.4737}$$

26.5263
16.5263
-23.4737
1.5263
1.5263
-18.4737
11.5263
-13.4737
-0.4737
14.5263
-18.4737
26.5263
16.5263
-23.4737
1.5263
1.5263
-18.4737

11.5263

-13.4737

The sum of the deviations is -0.0003, which except for round off, equals 0.

$$\sum (X - M) = 0$$

If each deviation is squared and the squared deviations are summed you get 4928.737. If this is divided by n – 1 you get 273.819. This is the variance of the freshmen weights. ***The variance is defined as***

$$S^2 = \sum (X - M)^2 / (n\text{-}1)$$

EXCEL has a ***built in function*** for variance, VAR(X). =VAR(F2:F20) gives 273.819.

If the expression for S^2 is square rooted you obtain the standard deviation or generally

The standard deviation is S =square root (S^2) = 16.5474

EXCEL has a ***built in function*** for standard deviation = STDEV(F2:F20) which gives 16.5474.

If the freshmen weights are entered into the MINITAB worksheet and the descriptive statistics N, S^2, and S are requested the following output is obtained.

Descriptive Statistics: wt

```
         Total
Variable  Count  StDev  Variance
wt          19   16.55   273.82
```

If freshmen weights are entered into the STATISTIX worksheet and the descriptive statistics N, S^2, and S are requested the following output is obtained.

Statistix 9.0   2/16/2013, 12:50:29 PM

Descriptive Statistics

Variable	N	SD	Variance
wt	19	16.547	273.82

If freshmen weights are entered into the SPSS worksheet and the descriptive statistics N, S^2, and S are requested the following output is obtained.

Descriptive Statistics

	N	Std. Deviation	Variance
freshmenwts	19	16.54747	273.819
Valid N (listwise)	19		

The following terms have been introduced in chapter 3: ***Range = Max – Min. The deviation of a variable from the mean is defined as***

$$(X - M).$$

$\sum(X - M) = 0$

The variance is

$$S^2 = \sum(X - M)^2 / (n\text{-}1)$$

The standard deviation is S = square root (S^2)

  A built in function is a function that is already built into a software package. A pull down menu is all that is required to execute the function. In EXCEL = is placed in front of the function and the function is executed.

Problems

Do 1 through 3 using EXCEL. Print your output for 1 through 3.

1. Find $\sum(X - M)$ where X represents internet times in the file spring13.xls.

2. Find $\sum(X - M)^2$ where X represents the internet times in the file spring13.xls.

3. Find $\sum(X - M)^2/(n - 1)$ where X represents internet times and n is the number of internet times.

4. Find the square root of the answer in problem 3.

5. Use the built in function $=var(X)$ in EXCEL and compare your answer with the answer in problem 3.

6. Use the built in function $= stdev(X)$ in EXCEL and compare your answer with the answer in problem 4.

Chapter 4

An experiment is any operation or procedure whose outcome cannot be predicted with certainty. The set of all possible outcomes for an experiment is called the sample space of the experiment. Software can be used to demonstrate probability.

For the experiment of tossing 2 dice, the sample space can be represented in EXCEL as follows:

1,6	2,6	3,6	4,6	5,6	6,6
1,5	2,5	3,5	4,5	5,5	6,5
1,4	2,4	3,4	4,4	5,4	6,4
1,3	2,3	3,3	4,3	5,3	6,3
1,2	2,2	3,2	4,2	5,2	6,2
1,1	2,1	3,1	4,1	5,1	6,1

An event is a subset of the sample space consisting of at least one outcome from the sample space. If the event consists of exactly one outcome, it is called a simple event. If an event consists of more than one outcome, it is called a compound event.

Probability is a measure of the likelihood of the occurrence of some event. The probability of any event E is represented by the symbol P(E). P(E) is a real number between 0 and 1.

$$0 \leq P(E) \leq 1$$

The probability that some outcome in the sample space will occur is 1.

$$P(S) = 1 \text{ or } P(E_1) + ... + (E_n) = 1$$

The classical definition of probability is appropriate when all outcomes of an experiment are equally likely. *For an event consisting of k outcomes, the classical probability of event E is P(E) = k/n where n is the total number of outcomes.*

The probability that the sum on the dice is 7 is $6/36 = 0.1667$, since the event that the sum on the dice is seven has 6 equally likely outcomes.

Another important concept is conditional probability. *If it is known that an event B has occurred, the conditional probability of event A is*
$$P(A \mid B) = P(A \text{ and } B) / P(B).$$

If event B in the above example is that die 1 is an odd number, the **reduced sample space** is

1,1	3,1	5,1
1,2	3,2	5,2
1,3	3,3	5,3
1,4	3,4	5,4
1,5	3,5	5,5
1,6	3,6	5,6

Suppose event A is that the sum of the dice is seven. $P(A \mid B) = 3/18 = 1/6 = P(A \text{ and } B) / P(B)$ where $P(A \text{ and } B)$ and $P(B)$ is computed from the sample space.

If $P(A \mid B) = P(A)$ then A and B are said to be independent events. If $P(A \mid B) \neq P(A)$, A and B are said to be dependent events, provided $P(B) > 0$.

Note that A and B are independent events in the above example.

From the equation for conditional probability $P(A \text{ and } B)$ can be written as
$$P(A \text{ and } B) = P(A \mid B) \, P(B)$$

Two or more events are said to be mutually exclusive if the events do not have any outcomes in common. They are events that cannot occur together. *If two events are mutually exclusive then*
$$P(A \text{ and } B) = 0.$$

To every event A, there corresponds another event A^c, called the complement of A, and consisting all other outcomes in the sample space not in event A. The following equation is true

$$P(A) + P(A^c) = 1$$

The **multiplication rule** is given by the formula $P(A \text{ and } B) = P(A) \, P(B \mid A)$. If A and B are independent events then
$$P(A \text{ and } B) = P(A) \, P(B).$$

The addition rule is $P(A \text{ or } B) = P(A) + P(B) - P(A \text{ and } B)$. If A and B are mutually exclusive,
$$P(A \text{ or } B) = P(A) + P(B).$$

A **Discrete random variable** can take on a countable number of values. The **probability distribution** of a Discrete random variable is the table of the values a discrete random variable can assume along with the probabilities the variable takes on these values.
Consider the example given previously of rolling a pair of dice. The sample space can be represented as follows in the EXCEL worksheet. Let X represent the sum of the two dice.

A	B	C	D	E	F
1	1	2	4	1	5
1	2	3	4	2	6
1	3	4	4	3	7
1	4	5	4	4	8
1	5	6	4	5	9
1	6	7	4	6	10
2	1	3	5	1	6
2	2	4	5	2	7
2	3	5	5	3	8
2	4	6	5	4	9
2	5	7	5	5	10
2	6	8	5	6	11
3	1	4	6	1	7
3	2	5	6	2	8
3	3	6	6	3	9
3	4	7	6	4	10
3	5	8	6	5	11
3	6	9	6	6	12

1 is entered into A1 and B1. =Sum(A1:B1) is entered into C1 and a click and drag is performed from C1 to C18. 4 is entered into D1 and 1 into E1. =Sum(D1:E1) is entered into F1 and a click and drag is performed from F1 to F18. The following probability distribution is found:

x	2	3	4	5	6	7	8	9	10	11	12
P(x)	1/36	2/36	3/36	4/36	5/36	6/36	5/36	4/36	3/36	2/36	1/36

Note the following properties of P(x): *P(x) $\geq$ 0 and $\sum P(x) = 1$. These are two properties that any probability distribution has.*

The mean of a discrete random variable is $\mu = \sum xP(x)$. The variance of a discrete random variable is $\sigma^2 = \sum x^2 P(x) - u^2$. The standard deviation is the square root of the variance.

Suppose we find μ and σ from an EXCEL work sheet as follows.

A	B	C	D
x	p(x)	x*P(x)	x^2*P(x)
2	0.027778	0.055556	0.111111
3	0.055556	0.166667	0.5
4	0.083333	0.333333	1.333333

						Standard	
5	0.111111	0.555556	2.777778				
6	0.138889	0.833333	5				
7	0.166667	1.166667	8.166667				
8	0.138889	1.111111	8.888889				
9	0.111111	1	9				
10	0.083333	0.833333	8.333333				
11	0.055556	0.611111	6.722222				
12	0.027778	0.333333	4				
sum		7	54.83333	variance	5.83333	dev	2.415229

We find the mean μ to be 7 and the standard deviation to be $\sigma = 2.415$. ***These are called the population mean and the population standard deviation.***

What is the connection between the sample mean, M, and the population mean μ? What is the connection between the sample variance, s^2, and the population variance, σ^2? In the above example, the population is all the casinos. The sample would be a sample from the casinos. The population of X has the population distribution:

x	2	3	4	5	6	7	8	9	10	11	12
P(x)	1/36	2/36	3/36	4/36	5/36	6/36	5/36	4/36	3/36	2/36	1/36

The population has mean $\mu = 7$ with standard deviation $\sigma = 2.415$. Suppose a random sample consisting of 20 rolls of a pair of dice have the outcomes: 7, 4, 5, 8, 5, 3, 4, 6, 8, 5, 2, 10, 7, 9, 9, 9, 9, 9, 8, 4, and 8. The sample mean is M = 6.5 and the sample standard deviation is S =2.351. If sample after sample of size 20 were obtained, we would find that the sample mean and sample standard deviation vary. However the population mean and standard deviation remain constant at 7 and 2.415.

The following terms have been introduced in chapter 4.

An experiment is any operation or procedure whose outcome cannot be predicted with certainty. The set of all possible outcomes for an experiment is called the sample space of the experiment.

An event is a subset of the sample space consisting of at least one outcome from the sample space. If the event consists of exactly one outcome, it is called a simple event. If an event consists of more than one outcome, it is called a compound event.

Probability is a measure of the likelihood of the occurrence of some event.

$$0 \leq P(E) \leq 1$$

$$P(S) = 1 \text{ or } P(E_1) + \ldots + (E_n) = 1$$

For an event consisting of k equally likely outcomes, the classical probability of event E is $P(E) = k/n$ where n is the total number of outcomes.

If it is known that an event B has occurred, the conditional probability of event A is

$$P(A \mid B) = P(A \text{ and } B) / P(B).$$

If $P(A \mid B) = P(A)$ then A and B are said to be independent events. If $P(A \mid B) \neq P(A)$, A and B are said to be dependent events.

.   *If two events are mutually exclusive then*

$$P(A \text{ and } B) = 0.$$

$$P(A) + P(A^c) = 1$$

The addition rule is $P(A \text{ or } B) = P(A) + P(B) - P(A \text{ and } B)$. If A and B are mutually exclusive,

$$P(A \text{ or } B) = P(A) + P(B).$$

$P(x) \geq 0$ and $\sum P(x) = 1$. These are two properties that any probability distribution has.

The mean of a discrete random variable is $\mu = \sum x P(x)$. The variance of a discrete random variable is $\sigma^2 = \sum x^2 P(x) - u^2$. The standard deviation is the square root of the variance.

Problems

The following problems apply to the experiment of tossing three dice.

1. Give the sample space in an EXCEL worksheet.

2. Give the complementary event for the event that the first and second die are each the number 1.

3. Event A is die 1 is a one. Event B is that die 3 is a three. Give the points in A and B. Find P(A and B).

4. For events A and B in problem 3 give the points in A or B. Find P(A or B).

5. Give the distribution of discrete random variable X, where X is the sum on the three dice.

6. Find the mean of the variable X from problem 5.

7. Find the standard deviation of the variable X from problem 5.

Chapter 5

Suppose we have an experiment consisting of n trials. On each trial one of two outcomes called **success** and **failure** can occur. The discrete random variable X counts the number of successes that occur in the n trials. X is a binomial random variable. The distribution of X is found in EXCEL and MINITAB as well as STATISTIX. A card is selected from a standard deck. Success is the event an Ace is selected. It is replaced and another is selected. This is repeated for 5 trials. The Binomial distribution of X can be found using EXCEL.

x	P(x)	cum
0	0.669898	0.669898
1	0.279426	0.949324
2	0.046622	0.995946
3	0.003889	0.999835
4	0.000162	0.999997
5	2.71E-06	1

Put the values, x, that the binomial variable can assume in column A. Enter = BINOMDIST(A2,5,0.077,0), where the probability of success is 4/52 or 0.077, under P(x) and perform a click-and-drag to obtain the probabilities of selecting an ace x number of times. Replace the 0 with 1 in the last parameter to obtain the cumulative probabilities. To get the mean μ, perform $\sum x\, P(x)$. To obtain the variance σ^2, find $\sum x^2 P(x) - \mu^2$. To find the standard deviation take the square root of the variance.

To find the same information using MINITAB do the following. Label C1 as x, label C2 as P(x) and C3 as Cum. Enter 0 through 5 under x. Give the pull down **Calc** $\rightarrow$ **Probablity Distribution** $\rightarrow$ **Binomial**. Fill in the dialog box as follows.

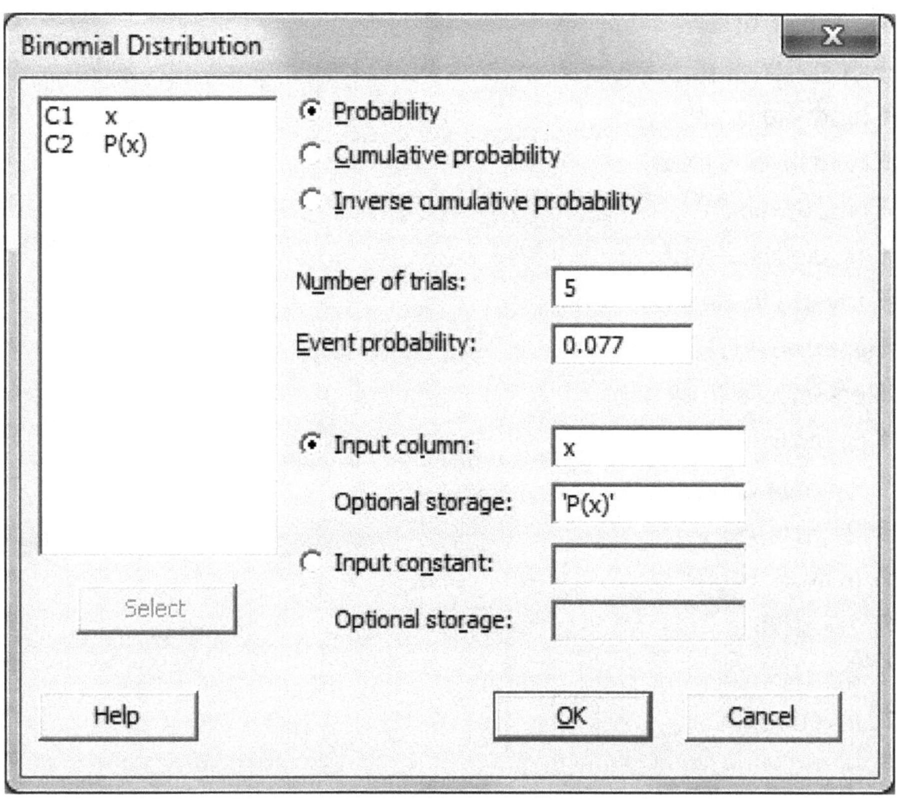

x	P(x)
0	0.669898
1	0.279426
2	0.046622
3	0.003889
4	0.000162
5	0.000003

The calculator can be used to show $\mu = 0.385$ and $\sigma = 0.596$.

STATISTIX gives P(0) as follows

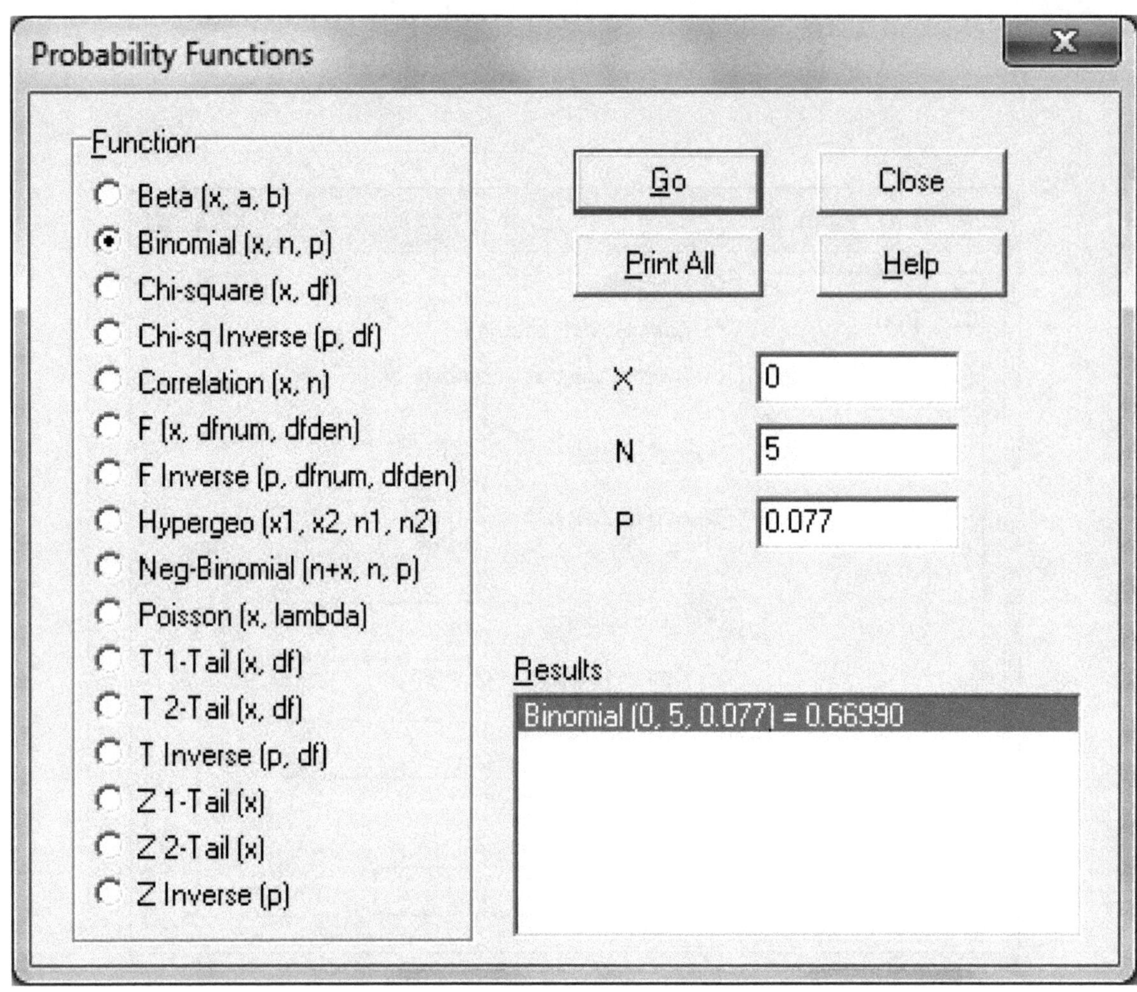

This could be used to find P(1), P(2), P(3), P(4), and P(5) also.

It is clear that software can be used to obtain the binomial distribution.

The following is taken from the Internet.

Handedness

From Wikipedia, the free encyclopedia

Jump to: navigation, search

This article is about left and right handedness in humans. For other uses, see Handedness (disambiguation).

"Handed" redirects here. For other uses, see Handed (disambiguation).

"Right Hand" redirects here. For other uses, see Right Hand (disambiguation).

"Southpaw" redirects here. For boxing stance, see Southpaw stance. For other uses, see Southpaw (disambiguation).

Handedness is a vague term that does not have a fixed or agreed definition. Although the terms *left* and *right* are enough to define handedness in an ordinary disclosure, they do not suffice for scientific research. [1] The scientific definition of handedness can be based on two theoretical assumptions. It can be defined as the hand that performs faster or more precisely on tasks or the hand that one prefers to use, regardless of performance. [2] In a scientific study, it should be recognized that handedness is not a discrete variable (right or left), but a continuous one that can be expressed at various levels between strong left and strong right. [1] There are four different types of handedness that include: left-handedness, right-handedness, mixed-handedness, and ambidexterity. [2]

- *Right-handedness* is most common. Right-handed people are more dexterous with their right hands when performing tasks. A variety of studies suggest that 70–90% of the world population is right-handed. [3][4]
- *Left-handedness* is less common than right-handedness. Left-handed people are more dexterous with their left hands when performing tasks. A variety of studies suggest that approximately 10% of the world population is left-handed. [5]
- *Mixed-Handedness* is the change of hand preference between different tasks. This is common in the population with about a 30% prevalence. [1]
- *Ambidexterity* is exceptionally rare, although it can be learned. A truly ambidextrous person is able to do any task equally well with either hand. Those who learn it still tend to favor their originally dominant hand. [2]

Suppose it is accepted that 10 percent of the population is left-handed. If you choose 10 people randomly, answer the following probability problems. Use MINITAB, give the probability distribution of X = the number of left-handers you might find in the 10. Also give the cumulative distribution of X.

x	P(x)	Cum
0	0.348678	0.34868
1	0.387420	0.73610
2	0.193710	0.92981
3	0.057396	0.98720
4	0.011160	0.99837
5	0.001488	0.99985
6	0.000138	0.99999
7	0.000009	1.00000
8	0.000000	1.00000
9	0.000000	1.00000
10	0.000000	1.00000

The most likely number of left-handers is 1. The probability of at most 2 left-handers is the sum of P(x) for x equal to zero, one, and two which is equal to 0.92981. The probability of at least 2 is $1 - 0.73610 = 0.2639$ (1 – prob at most 1). The mean μ equals 1. The standard deviation is 0.948.

Consider the heights of adult males. They have a mean of 70 inches and a standard deviation of 3 inches. Their normal distribution is given by the following pull down in MINITAB **Graph → Probability distribution plot**. View Single is selected in the dialog box.

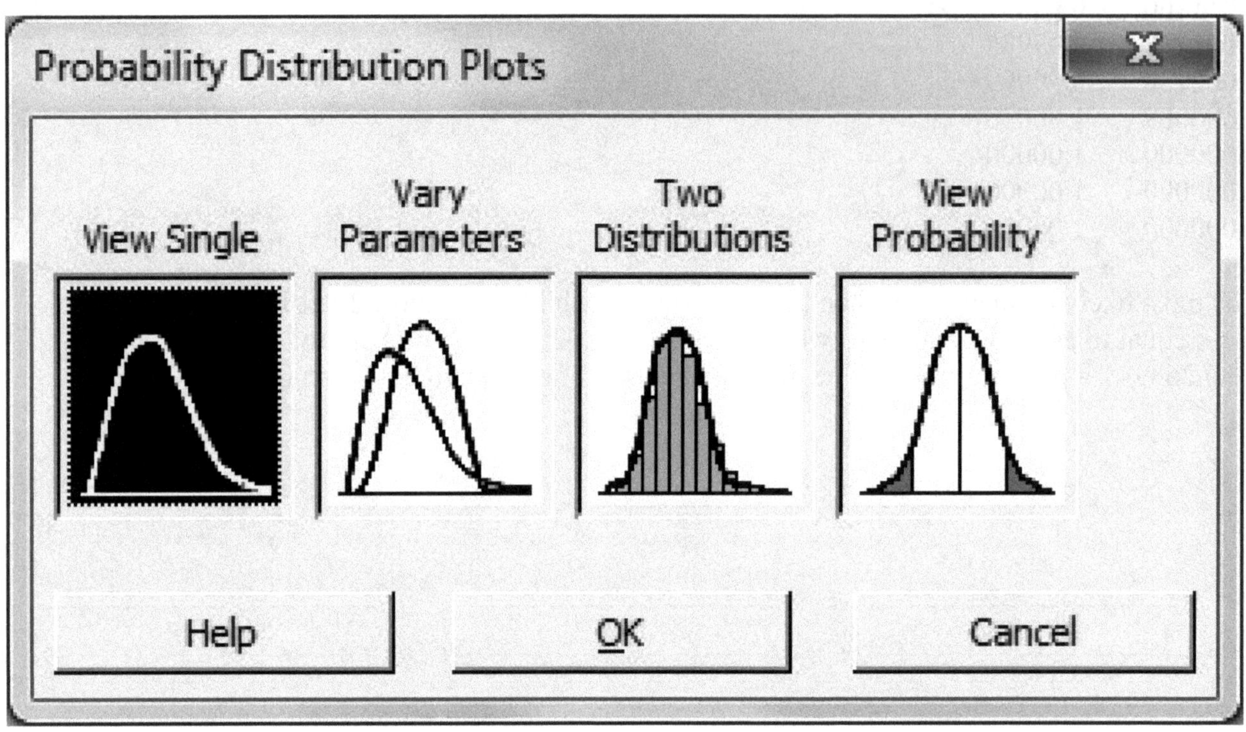

In the next dialog box choose normal distribution with mean 70 and standard deviation 3. Ths following graphic is produced.

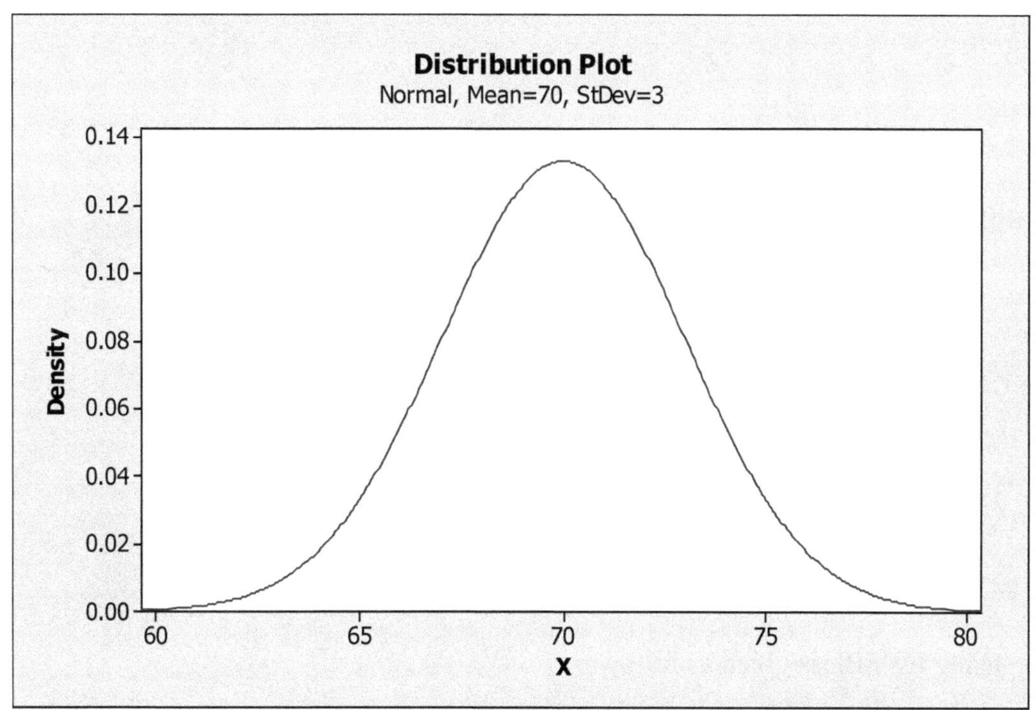

To find the percent of males that are 73 inches or shorter, we need to find the cumulative area that are 73 or less. The pull down **Calc → Probability distributions → Normal** gives the following dialog box which is filled as shown

This produces the following output

Cumulative Distribution Function

```
Normal with mean = 70 and standard deviation = 3

  x   P( X <= x )
 73      0.841345
```

We see that 84.13 percent are 73 inches or less.

This solution using EXCEL is given as follows.

=NORMDIST(73,70,3,1) which yields 0.841345 or 84.13%

To find the percent between 69 and 72 using MINITAB:

Cumulative Distribution Function

```
Normal with mean = 70 and standard deviation = 3

  x   P( X <= x )
 72      0.747507
```

Cumulative Distribution Function

```
Normal with mean = 70 and standard deviation = 3

  x   P( X <= x )
 69      0.369441
```

The percent between 69 and 72 inches is 0.747507-0.369441 = 0.378066 or 37.8%.

The solution using EXCEL is =NORMDIST(72,70,3,1) - NORMDIST(69,70,3,1) which equals 0.378066. To find the percent that are taller than 6 foot (72 inches) using MINITAB find the percent shorter than 72 inches as follows.

Cumulative Distribution Function

```
Normal with mean = 70 and standard deviation = 3

  x   P( X <= x )
 72      0.747507
```

This output states that 74.755% are 72 inches or shorter. The complement of this is 72 inches or taller. The P(X $\geq$ 72 inches) = 1 – P(X $\leq$ 72 inches) = 1 – 0.7475 = 0.2525 or 25.25 % are 6 feet or taller. To find this using EXCEL, Find =NORMDIST(72,70,3,1) which equals 0.7475. Subtract this from 1 to get 1 – 0.7475 = 0.2525 or 25.25%. or, give the EXCEL command =1-NORMDIST(72,70,3,1) = 0.2525.

The software can be used to find percentiles of a normal distribution. Suppose we wish to find the 90th percentile of male heights if it is known that male heights are normally distributed with mean equal to 70 inches and standard deviation equal to 3 inches. We wish to know the male height such that 90% of the males have a height which is less than that. The symbol for the 90th percentile is P_{90}. Use the same MINITAB pull down that you have used previously. **Calc $\rightarrow$ probability distribution $\rightarrow$ normal**. Fill in the dialog box as follows. Fill in the mean at 70, the standard deviation of 3, select inverse cumulative probability, and input constant at 0.90 for the 90th percentile. This is shown in the following dialog box.

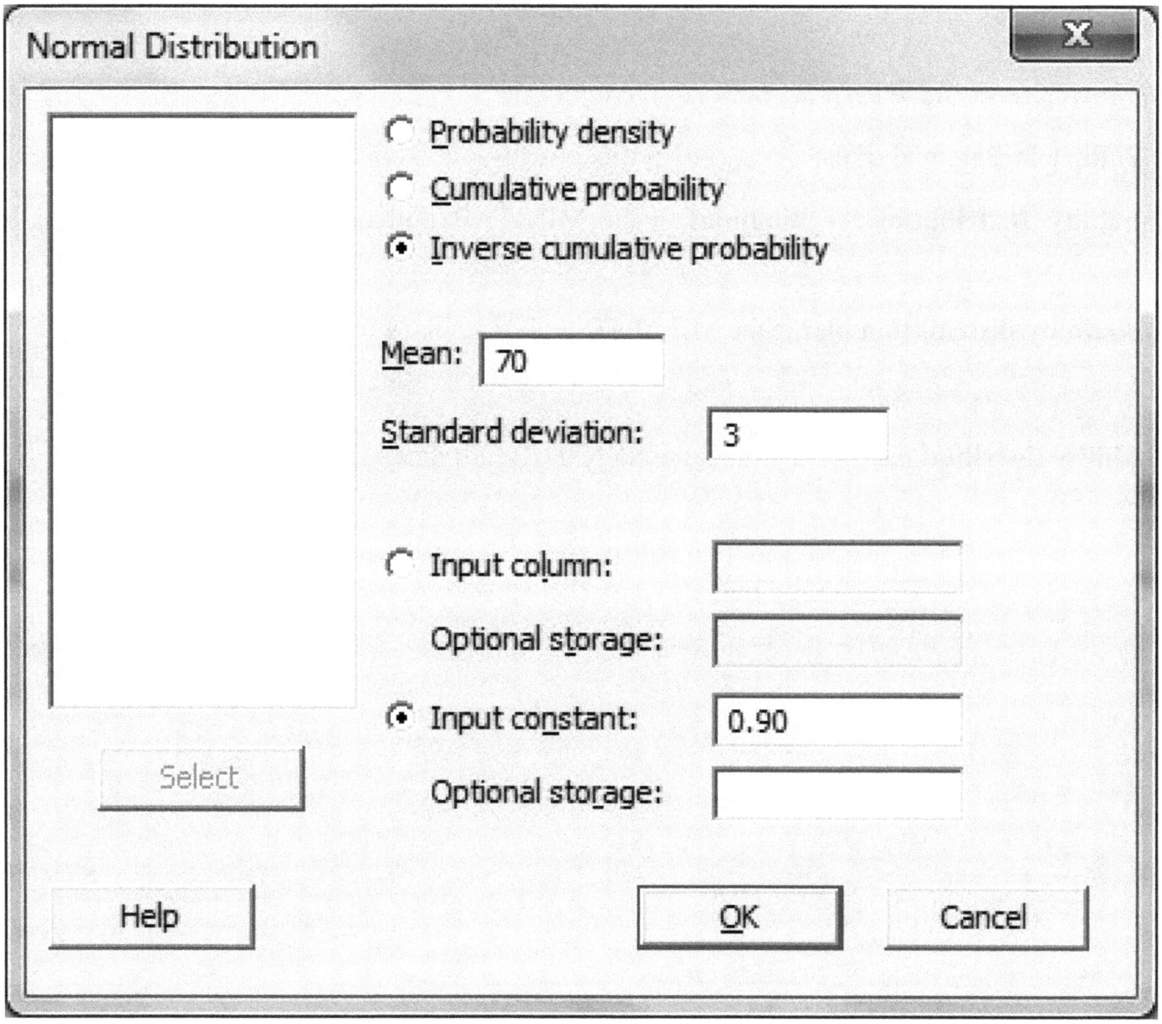

The following output is produced.

Inverse Cumulative Distribution Function
```
Normal with mean = 70 and standard deviation = 3

P( X <= x )         x
        0.9  73.8447
```
$P_{90} = 73.845$

Ninety percent of the males are shorter than 73.845 inches.

To find this in EXCEL give the command =NORMINV(0.9,70,3) and the answer 73.845 is returned.

To find the first quartile using EXCEL, give the command =NORMINV(0.25,70,3) which gives 67.977
inches.

The following terms have been introduced in chapter 5:

Success and Failure in Binomial trials.

Calc → Probablity Distribution → Binomial is the MINITAB pull down to give the Binomial
distribution.

Graph → Probability distribution plot is the MINITAB pull down used to draw the normal curve.

Calc → Probability distributions → Normal is the MINITAB pull down used to solve normal
probability problems.

Problems

1. Ten trials are performed in which a card is selected on each trial. If the card is a club a success occurs If any other card is obtained a failure occurred. The card is replaced before another is drawn. X represents number of clubs selected. Give the values that X can assume, the binomial probabilities and the cumulative binomial probabilities. Print your output.

2. Refer to problem 1. Find the probability that at most 3 clubs occur in the 10 trials.

3. Refer to problem 1. Find the probability that at least 2 clubs occur in the 10 trials.

4. Refer to problem 1. Find the value of μ in problem 1.

5. Refer to problem 1. Find the value of σ in problem 1.

6. Adult females have weights that are normally distributed with mean equal to 150 pounds and standard deviation equal to 25 pounds. Let X represent the female weights. Print your output for each part.

   a Find the probability X is less than 165 pounds.

   b Find the probability that X is greater than 140 pounds.

   c Find the probability that X is between 160 pounds and 170 pounds.

   d Find the value of Q_3, the third quartile of female weights.

Chapter 6

In order to obtain information about some population, either a *census* of the whole population is taken or a *sample* is chosen from the population and the information is inferred from the sample. The second approach is usually taken, since it is much cheaper to obtain a sample than to conduct a census. In choosing a sample, it is desirable to obtain one that is representative of the population. The average weight of the football players at a college would not be a representative estimate of the average weight of students attending the college, for example. A *simple random sample* of size n from a population of size N is one selected in such a way that every sample of size n has the same chance of occurring. In *simple random sampling with replacement,* a member of the population can be selected more than once. In *simple random sampling without replacement,* a member of the population can be selected at most once. Simple random sampling without replacement is the most common type of simple random sampling.

Suppose a table contains the 2000 most active New York Stock Exchange issues. We can use MINITAB to select 25 of these Exchange issues. Use the pull down **Calc → Random data →Integer**. The following dialog box is completed as shown.

The following output is produced.

```
MTB > print c1
```

Data Display

```
C1
   1667     238    1940     141    1926     276     281     400    1576    1384     128
    353     990     797    1891     564    1419     940    1117     604    1906    1688
     77     153    1473
```

The stocks are numbered from 1 to 2000. The above stocks are selected.

Suppose EXCEL is used to select the sample of 25.

Enter **=RANDBETWEEN**(1,2000) in any empty cell and click and drag for 25 cells. The
following is obtained.

```
1445
 100
 576
 744
1788
1754
 388
1472
1409
 995
1910
1024
1544
 420
1046
 114
1378
 903
1510
 799
1846
1905
 949
1686
1469
```

This is your random sample of 25 numbers between 1 and 2000.

Suppose Midwestern University has 20,000 students attending and suppose 25% are taking 6 credit hours, 25 % are taking 9 credit hours, 25% are taking 12 credit hours, and 25% are taking 15 credit hours. If X is the number of credit hours taken by a student at Midwestern, the distribution of X is

x	6	9	12	15
P(x)	0.25	0.25	0.25	0.25

The mean of X is $\sum xP(x) = 10.5$ and the variance of X is $\sigma^2 = \sum x^2 p(x) - \mu^2 = 121.5 - 110.25 = 11.25$

Consider all samples of size n = 2. The distribution of Xbar is derived in the following EXCEL worksheet.

item 1	item2	mean	probability	xbar	P(xbar)	F*G	(F^2)*G
6	6	6	0.0625	6	0.0625	0.375	2.25
6	9	7.5	0.0625	7.5	0.125	0.9375	7.03125
6	12	9	0.0625	9	0.1875	1.6875	15.1875
6	15	10.5	0.0625	10.5	0.25	2.625	27.5625
9	6	7.5	0.0625	12	0.1875	2.25	27
9	9	9	0.0625	13.5	0.125	1.6875	22.78125
9	12	10.5	0.0625	15	0.0625	0.9375	14.0625
9	15	12	0.0625	μ(xbar)		10.5	115.875
12	6	9	0.0625	var(xbar)			5.625
12	9	10.5	0.0625	σ(xbar)			2.371708
12	12	12	0.0625				
12	15	13.5	0.0625				
15	6	10.5	0.0625				
15	9	12	0.0625				
15	12	13.5	0.0625				
15	15	15	0.0625				

The distribution of Xbar is derived above. The mean of Xbar is found to be μ, and the variance of Xbar is found to be $\sigma^2/2$.

Now, consider all samples of size n =3 selected from Midwestern and we wish to build the distribution of Xbar the mean of samples of size 3.

item1	item2	item3	mean	probability	Xbar	P(Xbar)	F*G	F^2*G
6	6	6	6	0.015625	6	0.015625	0.09375	0.5625
6	6	9	7	0.015625	7	0.046875	0.328125	2.296875
6	6	12	8	0.015625	8	0.09375	0.75	6
6	6	15	9	0.015625	9	0.15625	1.40625	12.65625

6	9	6	7	0.015625		10	0.1875	1.875	18.75
6	9	9	8	0.015625		11	0.1875	2.0625	22.6875
6	9	12	9	0.015625		12	0.15625	1.875	22.5
6	9	15	10	0.015625		13	0.09375	1.21875	15.84375
6	12	6	8	0.015625		14	0.046875	0.65625	9.1875
6	12	9	9	0.015625		15	0.015625	0.234375	3.515625
6	12	12	10	0.015625		μ(Xbar)		10.5	114
6	12	15	11	0.015625		var(xbar)			3.75
6	15	6	9	0.015625		σ(xbar)			1.936492
6	15	9	10	0.015625					
6	15	12	11	0.015625					
6	15	15	12	0.015625					
9	6	6	7	0.015625					
9	6	9	8	0.015625					
9	6	12	9	0.015625					
9	6	15	10	0.015625					
9	9	6	8	0.015625					
9	9	9	9	0.015625					
9	9	12	10	0.015625					
9	9	15	11	0.015625					
9	12	6	9	0.015625					
9	12	9	10	0.015625					
9	12	12	11	0.015625					
9	12	15	12	0.015625					
9	15	6	10	0.015625					
9	15	9	11	0.015625					
9	15	12	12	0.015625					
9	15	15	13	0.015625					
12	6	6	8	0.015625					
12	6	9	9	0.015625					
12	6	12	10	0.015625					
12	6	15	11	0.015625					
12	9	6	9	0.015625					
12	9	9	10	0.015625					
12	9	12	11	0.015625					
12	9	15	12	0.015625					
12	12	6	10	0.015625					
12	12	9	11	0.015625					
12	12	12	12	0.015625					
12	12	15	13	0.015625					
12	15	6	11	0.015625					
12	15	9	12	0.015625					

12	15	12	13	0.015625
12	15	15	14	0.015625
15	6	6	9	0.015625
15	6	9	10	0.015625
15	6	12	11	0.015625
15	6	15	12	0.015625
15	9	6	10	0.015625
15	9	9	11	0.015625
15	9	12	12	0.015625
15	9	15	13	0.015625
15	12	6	11	0.015625
15	12	9	12	0.015625
15	12	12	13	0.015625
15	12	15	14	0.015625
15	15	6	12	0.015625
15	15	9	13	0.015625
15	15	12	14	0.015625
15	15	15	15	0.015625

The distribution of Xbar is derived above. The mean of Xbar is found be μ, and the variance of Xbar is found to be $\sigma^2/3$. If this is continued one more step to samples of size n = 4, we find the following distribution for Xbar.

Xbar	P(Xbar)
6.00	0.003906
6.75	0.015625
7.50	0.039063
8.25	0.078125
9.00	0.121095
9.75	0.156240
10.50	0.171864
11.25	0.156240
12.00	0.121095
12.75	0.078125
13.50	0.039063
14.25	0.015625
15.00	0.003906

And it is found that $\mu(Xbar) = \mu$ and variance(Xbar) $= \sigma^2/4$. One other thing that we will do is look at the distribution of Xbar for n =2, 3, and 4. Scatter plots are shown for n = 2, 3, and 4.

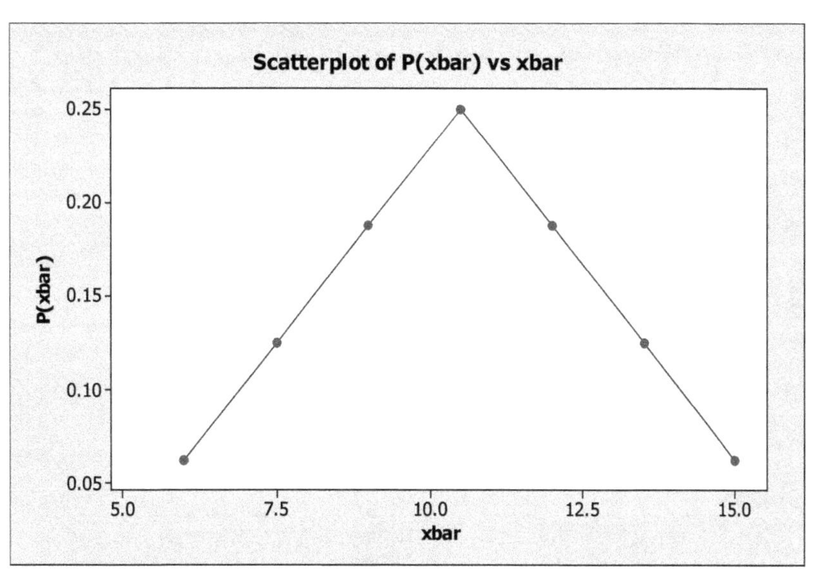

Scatterplot of P(xbar) vs xbar

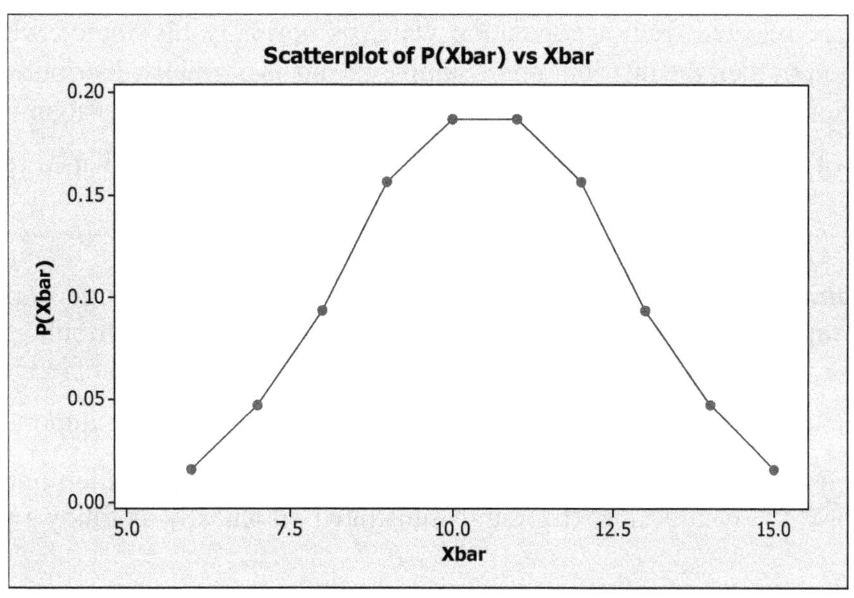

Scatterplot of P(Xbar) vs Xbar

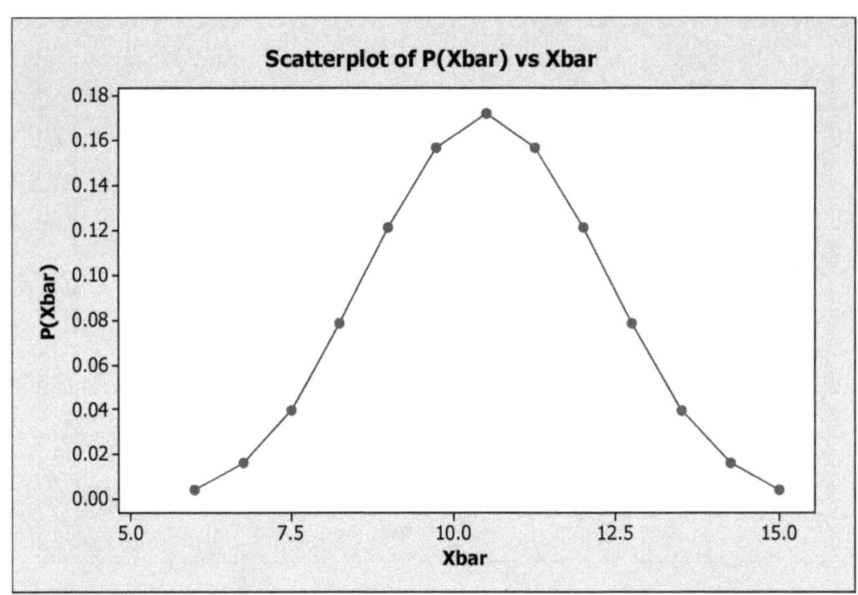

Note that as the sample size increases the curve approaches a normal curve.

If samples are selected from a population which is normally distributed with mean μ and standard deviation σ, then the distribution of sample means is normally distributed and the mean of this distribution is $\mu(Xbar) = \mu$, and the standard deviation of this distribution is $\sigma(xbar) = \frac{\sigma}{\sqrt{n}}$. The shape of the distribution of the sample means is normal or bell-shaped regardless of the sample size.

The ***central limit theorem*** states that when sampling from a large population of any distributional shape, the sample mean has approximately a normal distribution whenever the sample size is 30 or more. Furthermore, the mean of the distribution of sample means is $\mu(Xbar) = \mu$, and the standard deviation of this distribution is $\sigma(xbar) = \frac{\sigma}{\sqrt{n}}$. It is important to note that when sampling from a nonnormal distribution, xbar has a normal distribution only if the sample size is 30 or more. The central limit theorem is illustrated graphically as follows when n is 30 or more.

f(xbar)

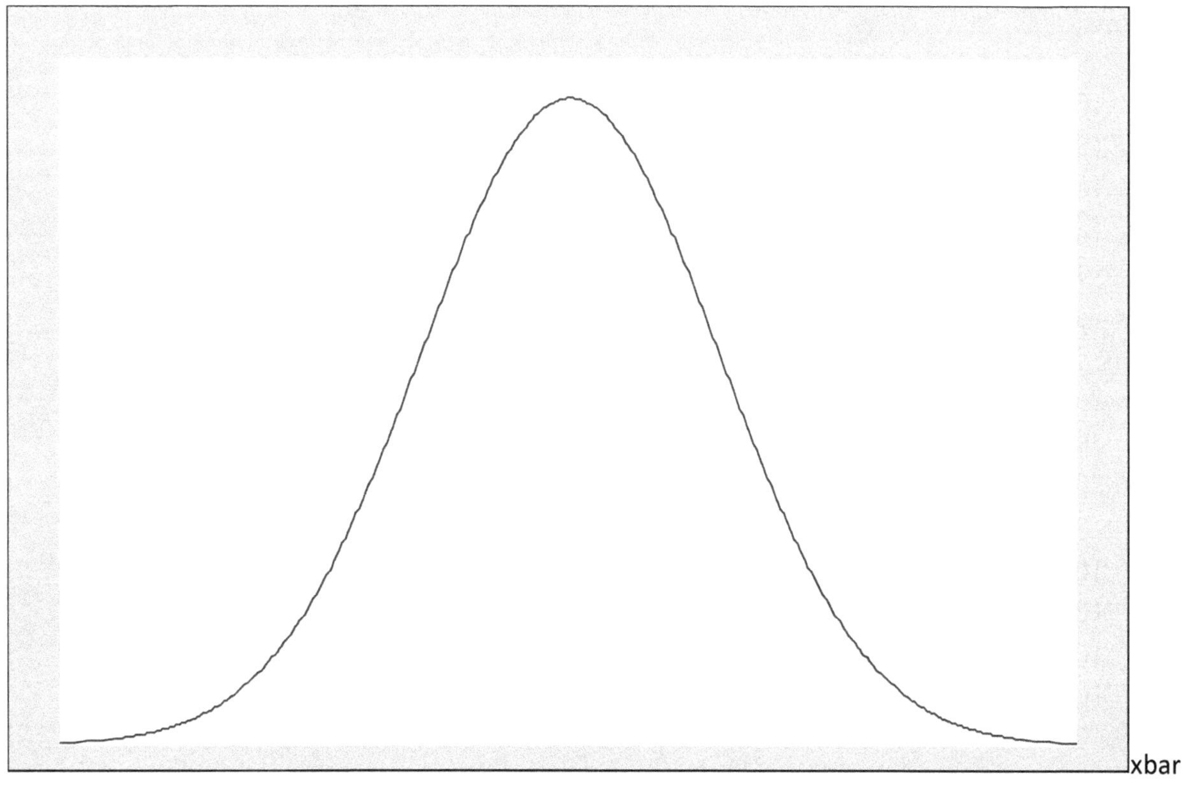

The above graphic illustrates that for samples greater than or equal to 30, Xbar has a distribution that is bell-shaped and centers at μ. The spread of the curve is determined by $\sigma(xbar) = \sigma$ / square root(n).

The transformed variable $Z = (Xbar - \mu) / \sigma(xbar)$ has a standard normal distribution, that is, a normal distribution with mean equal to 0 and standard deviation equal to 1. The following is also true $P(-1.96 < Z < 1.96) = 0.95$. From the past relationships we have:

$$P(-1.96 < (Xbar - \mu) / \sigma(xbar) < 1.96) = 0.95$$

When the inequality inside the probability statement is solved for μ, we have the following

The interval (Xbar – 1.96 σ(xbar), Xbar + 1.96 σ(xbar)) is called a 95% confidence interval for μ.

Suppose we enter the data in an EXCEL worksheet and we wished to find a 95% confidence interval for the mean time females who take Stat 3000 spend on the internet. Considering the data file spring13.xls as a sample of students who take Stat 3000. The time that the females spend on the internet in hours per week is:

feinternet

9	Xbar	13.4444
10	S	9.57092
10	S/sqrt(72)	1.12794
10		
10		
7		
5		
7		
20		
40		
20		
9		
10		
10		
10		
10		
7		
5		
7		
24		
10		
4		
4		
20		
24		
10		
3		
6		
12		
20		
5		
4		
4		
10		
3		

6
12
20
5
4
4
25
39
6
10
8
28
8
18
20
32
20
16
25
39
9
10
25
16
2
10
10
10
32
30
25
3
9
10
25
16
2

The sample mean is Xbar = 13.4444. Since σ is unknown, estimate it by S = 9.57092 and estimate σ(Xbar) by S/sqrt(72) = 1.12794.The estimated value of σ(xbar) is 1.12794.

Our estimated 95% confidence interval is *(xbar – 1.96 σ(xbar), xbar + 1.96 σ(xbar))*

or (13.4444 – 1.96*1.12794, 13.4444 + 1.96*1.12794) or (11.2334, 15.6552).

If we use MINITAB to find the 95% confidence interval, give the pull down menu **Stat →Basic Statistics → 1-sample Z**. The data for female internet times are entered in the work sheet. The following dialog box is filled out as shown. 95% confidence is entered into options.

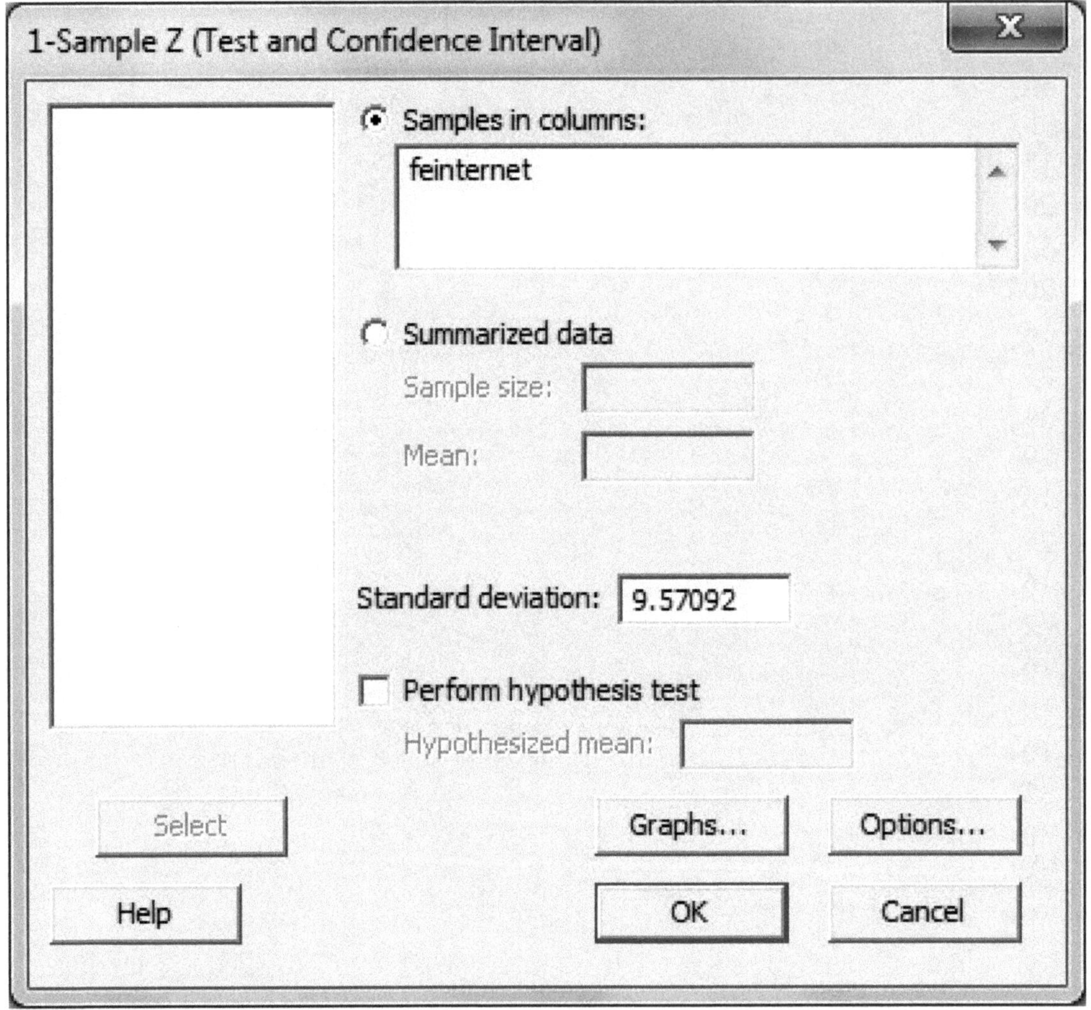

The following output is given.
One-Sample Z: feinternet

```
The assumed standard deviation = 9.57092


Variable      N    Mean   StDev  SE Mean       95% CI
feinternet   72   13.44    9.57     1.13   (11.23, 15.66)
```

This is the same output obtained as with EXCEL.

When a small sample (n < 30) is taken from a normally distributed population and the population standard deviation, σ, is unknown, a confidence interval for the population mean, μ, is given by

$$\text{Xbar} - t(S/\text{sqrt}(n)) < \mu < \text{Xbar} + t(S/\text{sqrt}(n))$$

Where t is from the t-distribution having (n-1) degrees of freedom.

Find a 95% confidence interval for the population mean of the age of freshmen students in Stat 3000. Use as your sample the freshmen in the file spring13.xls.Use MINITAB first. Put the freshmen ages in the MINITAB worksheet.

<u>Freshmen ages</u>

19
18
18
18
18
19
18
18
18
18
19
19
18
18
18
18
19
18
18

Give the pull down **Stat → Basic Statistics → 1-sample t.** The following output is produced.

One-Sample T: freshmenages

```
Variable        N    Mean   StDev  SE Mean        95% CI
freshmenages   19  18.263   0.452    0.104  (18.045, 18.481)
```

Next use SPSS to find a 95% confidence interval for μ. Enter the data in the Data Editor. Give the pull down **Analyze → Compare means → one sample t test** The output is as follows:

One-Sample Statistics

	N	Mean	Std. Deviation	Std. Error Mean
freshmenages	19	18.2632	.45241	.10379

One-Sample Test

	Test Value = 0					
	t	df	Sig. (2-tailed)	Mean Difference	95% Confidence Interval of the Difference	
					Lower	Upper
freshmenages	175.961	18	.000	18.26316	18.0451	18.4812

The same 95% confidence interval is produced as in MINITAB.

The following terms from statistical methods are introduced in this chapter.

*Census   sample   simple random sample   simple random sampling with replacement   simple random sampling without replacement   **Calc → Random data →Integer**   **=RANDBETWEEN(minimum, maximum)**   central limit theorem The interval (Xbar – 1.96 σ(xbar), Xbar + 1.96 σ(xbar)) is called a 95% confidence interval for μ   **Stat →Basic Statistics → 1-sample Z**   **Stat → Basic Statistics → 1-sample t***

Problems

1. Select a sample of 10 from the 50 states plus the district of Columbia.

1. Alabama	18. Kentucky	35. North Dakota
2. Alaska	19. Louisiana	36. Ohio
3. Arizona	20. Maine	37. Oklahoma
4. Arkansas	21. Maryland	38. Oregon
5. California	22. Massachusetts	39. Pennsylvania
6. Colorado	23. Michigan	40. Rhode Island
7. Connecticut	24. Minnesota	41. South Carolina
8. Delaware	25. Mississippi	42. South Dakota
9. District of Columbia	26. Missouri	43. Tennessee
10. Florida	27. Montana	44. Texas
11. Georgia	28. Nebraska	45. Utah
12. Hawaii	29. Nevada	46. Vermont
13. Idaho	30. New Hampshire	47. Virginia
14. Illinois	31. New Jersey	48. Washington
15. Indiana	32. New Mexico	49. West Virginia
16. Iowa	33. New York	50. Wisconsin
17. Kansas	34. North Carolina	51. Wyoming

Give the names of the 10 states that you select. Did you get the random numbers from EXCEL or MINITAB?

2. 1/3 of Students at Midwestern university take 6 credit hours, 1/3 take 9 credit hours and 1/3 take 12 credit hours. X represents the credit hours a student at Midwestern takes. Give the probability distribution of X. Find μ and σ for X. The number of students at Midwestern is large.

3. Give the probability distribution for Xbar for samples of size n = 2. Find μ(Xbar) and σ(Xbar).

4. Give the probability distribution for Xbar for samples of size n = 3. Find μ(Xbar) and σ(Xbar).

5. Find a 95% confidence interval for the mean time per week spent by females on the phone. Use the sample of female phone times from the file spring13.xls.

6. Find a 95% confidence interval for the mean time per week spent by males on the phone. Use the sample of male phone times from the file spring13.xls. Use SPSS and give your output.

Chapter 7

In this chapter, rather than estimate a population mean, we shall be interested in ***testing a hypothesis about a population mean***. Our research hypothesis will be of the form $\mu < a$, $\mu > a$, or $\mu \neq a$. Our null hypothesis will be of the form $\mu = a$. In testing hypothesis, two types of errors may occur.

	Null hypothesis is	
	True	False
Decision		
Do not reject null	correct decision	***Type II error***
Reject null	***Type I error***	correct decision

The first two letters of the Greek alphabet represent the size of the two errors.

$\alpha = P\{\text{rejecting the null hypothesis when it is true}\}$
$\beta = P\{\text{not rejecting the null when the null is false}\}$

Suppose we are interested in the average age of a college student taking their first statistics course. The hypothesis is that the average age is 25. This leads to the hypothesis that $\mu = 25$ versus $\mu \neq 25$. The sample contained in spring13.xls is a sample of 97 students in their first statistics course. Their age is one of 10 variables measured. The MINITAB descriptive statistics for age is

Descriptive Statistics: age

```
Variable   N   N*     Mean  SE Mean   StDev  Minimum       Q1  Median       Q3
age        97   0   23.629    0.702   6.915   18.000   20.000  21.000   25.000

Variable  Maximum
age        50.000
```

The sample mean is 23.629 years, the sample standard deviation is 6.915 years, and the sample standard error of their ages (S/square root of n) is 0.702 years. The ***test statistic*** for testing the hypothesis that $\mu = 25$ is $Z = (\text{Xbar} - \mu_0)/(\text{standard error of Xbar})$ or $(23.629 - 25)/0.702 = -1.95$
If we wanted the chance of making a type I error to be fixed at $\alpha = 0.05$, then we would have the following decision making procedure.

reject $\mu = 25$	do not reject $\mu = 25$	reject $\mu = 25$	
-1.96		1.96	Z

The value that the type I error is fixed at $\alpha = 0.05$ is called the ***significance level***. The significance level determines the ***critical values*** -1.96 and 1.96. The critical values determine the ***rejection region*** $\{Z < -1.96$ or $Z > 1.96\}$ and ***the non-rejection region*** $\{-1.96 < Z < 1.96\}$.

Since the test statistic is in the non-rejection region we do not reject the null that $\mu = 25$.

To illustrate how to find the critical values using MINITAB, give the pull down ***Calc →
probability distributions → Normal***. Fill in the dialog box as follows:

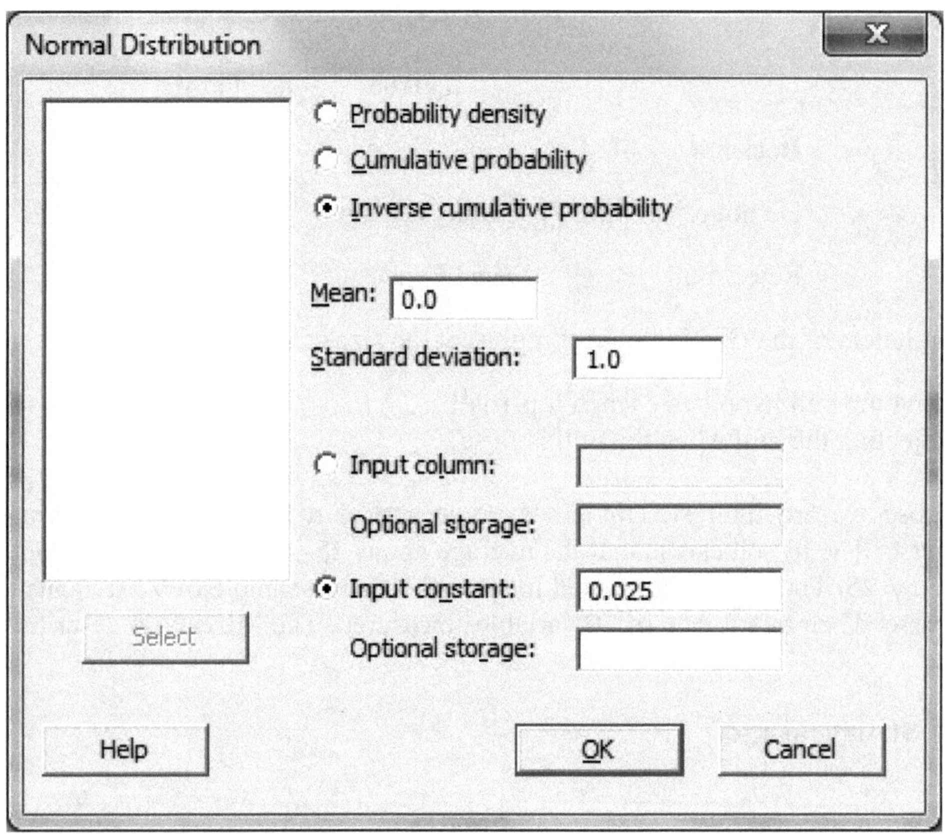

The output is as follows:

Inverse Cumulative Distribution Function

```
          Normal with mean = 0 and standard deviation = 1

        P( X <= x )           x        0.025   -1.95996
```

Because of the symmetry of the normal curve, the other critical value is 1.95996.

reject $\mu =25$	do not reject $\mu = 25$	reject $\mu= 25$	Z
critical value -1.96	critical value 1.96		

This test is performed for you in MINITAB. Read the data file into MINITAB and get the following.

gender	class	age	gpa	ht	wt	internet	phone	exercise	party
f	f	19	3.00	67	160	9	4	30	other
f	f	18	3.20	67	150	10	7	240	other
f	f	18	4.00	62	110	10	5	120	other
f	f	18	3.90	68	135	10	25	650	republican
f	f	18	3.80	67	135	10	15	150	republican
f	f	19	3.70	64	115	7	14	120	republican
f	f	18	3.50	74	145	5	2	240	other
f	f	18	3.00	68	120	7	8	120	republican
f	f	18	3.50	65	133	20	20	700	republican
f	f	18	4.00	64	148	40	1	60	other
f	f	19	3.40	64	115	20	15	300	republican
f	f	19	3.00	67	160	9	4	30	other
f	f	18	3.20	67	150	10	7	240	other
f	f	18	4.00	62	110	10	5	120	other
f	f	18	3.90	68	135	10	25	650	republican
f	f	18	3.80	67	135	10	15	150	republican
f	f	19	3.70	64	115	7	14	120	republican
f	f	18	3.50	74	145	5	2	240	other
f	f	18	3.00	68	120	7	8	120	republican
f	g	25	3.60	68	250	24	50	90	republican
f	g	48	3.00	63	200	10	7	270	democrat
f	g	44	4.00	67	168	4	2	120	other
f	g	50	3.80	65	159	4	3	90	democrat
f	g	27	3.00	67	160	20	10	30	other
f	g	25	3.60	68	250	24	50	90	republican
f	g	48	3.00	63	200	10	7	270	democrat
f	j	35	3.90	68	154	3	3	60	other
f	j	20	2.90	64	142	6	1	420	democrat
f	j	20	3.10	67	150	12	3	360	other
f	j	20	3.00	67	170	20	10	560	other
f	j	24	3.50	65	163	5	1	240	democrat
f	j	21	3.80	66	165	4	2	300	republican
f	j	21	3.60	65	125	4	3	0	republican
f	j	25	3.20	67	135	10	20	600	democrat
f	j	35	3.90	68	154	3	3	60	other
f	j	20	2.90	64	142	6	1	420	democrat
f	j	20	3.10	67	150	12	3	360	other
f	j	20	3.00	67	170	20	10	560	other
f	j	24	3.50	65	163	5	1	240	democrat
f	j	21	3.80	66	165	4	2	300	republican
f	j	21	3.60	65	125	4	3	0	republican
f	s	20	3.90	62	130	25	10	180	republican

f	s	21	2.80	63	190	39	1	120	republican
f	s	20	3.80	61	96	6	3	60	other
f	s	31	2.50	62	125	10	1	0	democrat
f	s	21	3.80	61	100	8	2	60	republican
f	s	20	2.60	67	160	28	15	0	other
f	s	20	3.70	63	145	8	1	180	other
f	s	20	3.00	63	150	18	2	120	other
f	s	20	3.50	70	125	20	20	210	other
f	s	43	3.70	67	280	32	21	185	democrat
f	s	20	2.90	63	140	20	2	500	other
f	s	24	2.70	66	150	16	3	180	other
f	s	20	3.90	62	130	25	10	180	republican
f	s	21	2.80	63	190	39	1	120	republican
f	sr	22	3.00	64	129	9	4	240	other
f	sr	21	3.90	66	160	10	10	120	democrat
f	sr	26	2.60	65	150	25	2	60	democrat
f	sr	23	3.40	67	135	16	8	360	republican
f	sr	27	2.90	65	135	2	2	60	democrat
f	sr	24	3.20	62	116	10	8	200	other
f	sr	20	2.90	69	139	10	15	240	democrat
f	sr	24	4.00	64	140	10	7	380	republican
f	sr	41	3.30	64	240	32	24	140	other
f	sr	21	3.30	61	120	30	30	60	republican
f	sr	21	3.75	67	160	25	10	100	other
f	sr	27	3.60	69	160	3	3	150	republican
f	sr	22	3.00	64	129	9	4	240	other
f	sr	21	3.90	66	160	10	10	120	democrat
f	sr	26	2.60	65	150	25	2	60	democrat
f	sr	23	3.40	67	135	16	8	360	republican
f	sr	27	2.90	65	135	2	2	60	democrat
m	j	22	3.50	75	190	20	1	1200	other
m	j	20	3.10	64	120	10	1	60	republican
m	j	22	3.40	74	185	10	5	250	other
m	j	22	3.20	71	109	38	4	50	other
m	j	20	3.20	72	150	6	14	240	republican
m	j	35	3.00	70	180	3	4	200	republican
m	j	20	3.30	72	270	5	2	360	other
m	j	28	2.90	69	172	5	4	500	other
m	j	24	3.30	66	140	5	2	300	democrat
m	s	20	4.00	72	180	17	2	450	democrat
m	s	20	3.00	70	165	10	2	700	republican
m	s	20	3.00	73	180	20	20	360	republican
m	s	23	3.10	75	175	21	2	150	republican
m	s	19	3.00	72	130	14	30	600	republican
m	sr	21	3.40	72	170	25	1	120	other
m	sr	23	3.20	72	180	16	8	240	other

m	sr	25	3.30	75	230	20		5	240		democrat

The remaining values in the 97 would follow.

Perform the pull down **Stat→ Basic statistics → 1 sample Z** and get the following output.

One-Sample Z: age

```
Test of mu = 25 vs not = 25
The assumed standard deviation = 6.915


Variable   N    Mean   StDev  SE Mean        95% CI          Z      P
age       97  23.629   6.915    0.702  (22.253, 25.005)  -1.95  0.051
```

This test is called the *Z-test* or the *large sample test* (n > 30) for the mean of a single population. It may be summarized as follows:

1: State the null and research hypothesis.

2: Use the standard normal distribution and the level of significance to determine the rejection region.

3: Compute the value of the test statistic as follows: $Z = (xbar - \mu_0) / (S/\text{square root}(n))$ where xbar is the mean of the sample, μ_0 is the population mean stated in the null hypothesis, S is the sample standard deviation, and n is the sample size.

4: State your conclusion.

A second way of conducting the test is to compare the *p-value* with the significance level. If the p-value is less than α, the significance level, reject the null hypothesis. If the p-value is not less than α, then do not reject the null hypothesis.

First we shall show how to compute the p-value. *The p-value is defined as P{Z < -1.95} + P{Z > 1.95}.*

```
Normal with mean = 0 and standard deviation = 1

     x   P( X <= x )
 -1.95     0.0255881
```

Due to the symmetry of the normal curve, we simply double this area to get the p-value. p-value = 2(0.025588) = 0.051176 or rounding to 3 decimal places, p-value = 0.051 and because this is larger than 0.05, do not reject μ = 25 years. Note that this is the same value given in the one-sample Z output.

The small sample test is used if n < 30. It may be summarized as follows:

1: State the null and the research hypothesis.

2: Use the t distribution, with degrees of freedom (n – 1) and level of significance α to determine the rejection region.

3: Compute the value of the test statistic. t = (xbar – μ_0)/(s / square root (n)), where xbar is the sample mean, μ_0 is the value of the population mean stated in the null hypothesis, s is the sample standard deviation, and n is the sample size.

Step 4: State your conclusion.

Suppose we wished to test that the average age of the graduate students taking the course is 25 years versus it is greater than 25, that is H_0: μ = 25 versus H_a : μ > 25. The significance level is α = 0.05. This is called an upper-tailed test. The null hypothesis will be rejected if t = (xbar – 25)/standard error is larger than the critical value. The critical value is the value such that only 0.05 of the area under the t curve with df = 6 is in the upper part of the curve. It is found as follows using MINITAB. Use the pull down menu *Calc → probability distributions → t.*

The descriptive statistics are

Descriptive Statistics: gradstudentage

```
Variable           N  N*    Mean  SE Mean  StDev  Minimum      Q1  Median      Q3
gradstudentage     7   0   38.14     4.47  11.82    25.00   25.00   44.00   48.00

Variable        Maximum
gradstudentage    50.00
```

t = (xbar – 25)/standard error = (38.14 – 25)/4.47 = 2.94. Since the computed test statistic is in the rejection region, reject the null hypothesis and accept μ > 25 years.

The critical t value is found as follows.

Inverse Cumulative Distribution Function

```
Student's t distribution with 6 DF

P( X <= x )       x
      0.95  1.94318
```

| Do not reject μ = 25 | Reject μ = 25 |

```
_________________________________________|_________________
                                     1.94318
```

Suppose we use the function *1 sample t*.

One-Sample T: gradstudentsage

```
Test of mu = 25 vs > 25
                                            95% Lower
Variable          N    Mean  StDev  SE Mean    Bound     T     P
gradstudentsage   7   38.14  11.82     4.47    29.46  2.94  0.013
```

The computed T value is 2.94. Therefore we reject μ = 25. Also the p-value is 0.013 and it is less than 0.05. If we use the p-value method of testing, we would also reject the null that μ = 25.

If the SPSS package is used, the output is:

One-Sample Statistics

	N	Mean	Std. Deviation	Std. Error Mean
gradstudentsages	7	38.1429	11.82411	4.46909

One-Sample Test

	Test Value = 25					
	t	df	Sig. (2-tailed)	Mean Difference	95% Confidence Interval of the Difference	
					Lower	Upper
gradstudentsages	2.941	6	.026	13.14286	2.2074	24.0783

This output corresponds with MINITAB. The two-tailed p-value is called Sig. (2-tailed) and is equal 0.026. The one-tailed p-value would be 0.013. The same as given by MINITAB.

If the STATISTIX software is used to do the test, the output is:

```
Statistix 9.0                                    3/30/2013, 3:53:48
PM

Hypothesis Test - One Mean

Null Hypothesis        mu = 25
Alternative Hyp        mu > 25
N                            7
Mean                    38.143
SD                      11.824

SE                      4.4691
DF                           6
T                         2.94
P                       0.0130
95% C.I. Lower Bound    27.207
95% C.I. Upper Bound    49.078
```

And the output agrees with MINITAB and SPSS.

The following terms from statistical methods are introduced in this chapter.

<table>
<tr><td></td><td colspan="2" align="center">*Null hypothesis is*</td></tr>
<tr><td></td><td align="center">*True*</td><td align="center">*False*</td></tr>
<tr><td>*Decision*</td><td></td><td></td></tr>
<tr><td>*Do not reject null*</td><td>*correct decision*</td><td>*Type II error*</td></tr>
<tr><td>*Reject null*</td><td>*Type I error*</td><td>*correct decision*</td></tr>
</table>

A test statistic is used for testing hypotheses.

The value that the type I error is fixed at α (usually $\alpha = 0.05$) is called the significance level. The significance level determines the critical values. The critical values determine the rejection region and the non-rejection region.

The MINITAB pull down Stat$\rightarrow$ Basic statistics $\rightarrow$ 1 sample Z is used for performing a test on μ.

A p-value is used for testing hypothesis. If the p-value is less than α, reject the null hypothesis. Otherwise, do not reject the null hypothesis.

The MINITAB pull down Stat$\rightarrow$ Basic statistics $\rightarrow$ 1 sample T is used for performing a test on μ.

Problems

1. Use the data in file spring13.xls to test that the mean time spent in exercise for students in a statistical methods class is 300 minutes per week versus it is not 300 minutes per week. Use significance level = 0.05. Use 1 sample Z of MINITAB. Give the output for 1 sample Z:exercise.What conclusion do you reach and why? What is the value of the test statistic? What is the p-value for the test?

2. Use the data of Spring13.xls to test that the mean age of seniors taking their statistical methods course is 25 years versus it is greater than 25 years at a level of significance of 0.01. Use MINITAB and give your output for one-sample t. What conclusion do you reach and why? What is the value of the test statistic? What is the p-value for the test?

3. Use the data of Spring13.xls to test that the mean exercise for freshmen taking their statistical methods course is 300 minutes versus it is not 300 minutes at a level of significance equal to 0.01. Give your SPSS output. What conclusion do you reach and why? What is the value of the test statistic? What is the p-value for the test?

4. Use the data of Spring13.xls to test that the mean age for females taking their statistical methods course is 23 years versus it is not 23 years at a level of significance equal to 0.01. Give your STATISTIX output. What conclusion do you reach and why? What is the value of the test statistic? What is the p-value for the test?

5. Use the data of Spring13.xls to test that the mean age for males taking their statistical methods course is 25 years versus it is not 25 years at a level of significance equal to 0.01. Give your STATISTIX output. What conclusion do you reach and why? What is the value of the test statistic? What is the p-value for the test?

Appendix

This chapter contains other data files formed from students taking a Statistical Methods course.

Fall11.xls

gender	class	age	gpa	ht	wt	internet	phone	exercise	party
f	j	21	3.4	64	105	7	3	90	republican
f	j	21	2.7	64	160	10	1	30	democrat
f	sr	21	3.8	60	90	28	2	60	republican
f	g	29	3.9	64	123	12	2	300	republican
f	g	25	3.3	62	120	12	8	300	other
f	sr	21	3.68	66	120	10	10	400	democrat
m	f	19	4	73	178	3	7	400	republican
f	sr	22	3	71	155	8	2	250	democrat
f	f	18	3.4	63	130	8	15	150	other
m	j	20	3.3	75	280	18	12	480	republican
f	j	24	2.9	63	195	16	2	250	other
m	j	20	2.7	74	150	24	2	200	republican
f	sr	23	3	67	155	10	2	500	other
m	j	24	2.4	72	205	3	2	300	democrat
m	j	22	2.5	70	190	24	10	270	democrat
f	g	38	3.9	64	125	16	2	120	democrat
f	sr	27	2.8	67	160	10	7	360	other
m	g	46	4	69	205	10	1	180	democrat
f	sr	20	3.2	62	115	5	6	240	other
f	sr	22	3.6	64	200	30	23	0	other
f	f	18	3.7	62	120	20	10	100	other
f	sr	19	3.2	68	135	30	50	180	other
m	sr	26	3.35	70	165	35	2	780	democrat
f	j	21	2.8	57	151	3	10	420	other
m	j	28	3.8	75	190	6	14	400	other
m	j	21	4	70	140	7	5	240	democrat
f	sr	19	3	66	134	4	10	120	other
m	s	20	3.8	65	135	14	1	0	democrat
f	s	20	2.5	60	90	12	18	840	democrat
f	j	21	3.2	64	140	9	20	240	other
m	j	20	3	75	160	12	10	60	other
f	f	20	3.9	67	180	7	7	60	democrat
f	j	23	4	63	140	6	10	180	other
m	s	19	3.2	76	235	18	5	400	republican
f	s	19	3.2	63	135	21	3	180	democrat
f	j	28	3.4	65	131	30	1	100	democrat
m	sr	23	2.7	76	205	40	5	480	other
f	j	22	3.2	66	142	10	5	300	republican
f	s	19	3.8	71	150	2	14	420	republican
f	j	21	3.4	64	120	7	5	240	republican

gender	class	age	gpa	Ht	Wt	internet	phone	exercise	party
f	j	20	3.5	64	127	3.6	10	840	democrat
f	s	20	3	60	96	14	5	30	democrat
f	s	20	3.2	67	135	28	32	360	other
f	j	20	3	67	170	25	10	100	republican
f	j	20	3.6	67	145	15	1	1800	democrat
m	sr	22	2.8	69	165	20	1	300	democrat
m	g	27	3	68	160	14	1	840	republican
m	j	26	3.46	70	264	10	0	0	democrat
m	g	26	3.6	73	195	25	3	240	democrat
m	s	20	3.2	71	205	1	3	240	republican
m	s	20	3.8	69	135	15	1	0	other
f	s	19	3.3	60	105	20	21	90	democrat
m	sr	22	3.5	72	160	24	6	240	democrat
f	f	18	3.7	69	135	30	50	120	republican
m	j	20	3.97	78	165	50	25	60	republican
f	s	19	3.7	66	165	10	2	400	republican
f	sr	21	3.6	65	121	3	5	180	other
f	s	19	3.5	62	120	10	10	300	democrat
f	g	35	3.2	70	170	7	6	250	other
f	s	19	3	65	170	5	20	300	democrat
f	sr	21	3.2	67	113	10	3	0	other
f	g	31	3.86	66	135	12	1	840	other
m	sr	23	2.6	72	179	21	14	300	republican
f	j	20	3.6	65	150	13	4	300	other
m	j	25	3.1	69	225	5	3	150	republican
m	sr	25	3.2	70	157	28	20	840	other
m	sr	21	3.4	71	185	12	13	120	other
m	s	19	3.8	66	175	20	2	0	other
m	s	20	3	69	160	4	6	420	other
m	j	24	3	75	180	12	5	150	republican

Spring12.xls

gender	class	age	gpa	Ht	Wt	internet	phone	exercise	party
f	s	20	3.9	62	130	25	10	180	republican
f	sr	22	3	64	129	9	4	240	other
f	j	35	3.9	68	154	3	3	60	other
f	j	20	2.9	64	142	6	1	420	democrat
f	s	21	2.8	63	190	39	1	120	republican
f	j	20	3.1	67	150	12	3	360	other
f	j	20	3	67	170	20	10	560	other
f	g	25	3.6	68	250	24	50	90	republican
f	sr	21	3.9	66	160	10	10	120	democrat
f	j	24	3.5	65	163	5	1	240	democrat
f	f	19	3	67	160	9	4	30	other
f	f	18	3.2	67	150	10	7	240	other
f	g	48	3	63	200	10	7	270	democrat

f	j	21	3.8	66	165	4	2	300	republican
f	sr	26	2.6	65	150	25	2	60	democrat
f	f	36	3.9	68	200	7	4	160	other
f	f	18	4	62	110	10	5	120	other
f	f	18	3.9	68	135	10	25	650	republican
f	f	18	3.8	67	135	10	15	150	republican
f	f	19	3.7	64	115	7	14	120	republican
f	j	21	3.6	65	125	4	3	0	republican
f	sr	23	3.4	67	135	16	8	360	republican
f	f	18	3.5	74	145	5	2	240	other
f	f	18	3	68	120	7	8	120	republican
f	sr	27	2.9	65	135	2	2	60	democrat
f	s	20	3.8	61	96	6	3	60	other
f	j	25	3.2	67	135	10	20	600	democrat
f	s	31	2.5	62	125	10	1	0	democrat
f	f	18	3.5	65	133	20	20	700	republican
f	f	18	4	64	148	40	1	60	other
f	s	21	3.8	61	100	8	2	60	republican
f	s	20	2.6	67	160	28	15	0	other
f	g	44	4	67	168	4	2	120	other
f	s	20	3.7	63	145	8	1	180	other
f	s	20	3	63	150	18	2	120	other
f	sr	24	3.2	62	116	10	8	200	other
f	s	20	3.5	70	125	20	20	210	other
f	s	43	3.7	67	280	32	21	185	democrat
m	s	20	4	72	180	17	2	450	democrat
m	j	22	3.5	75	190	20	1	1200	other
m	j	20	3.1	64	120	10	1	60	republican
m	j	22	3.4	74	185	10	5	250	other
m	s	20	3	70	165	10	2	700	republican
m	j	22	3.2	71	109	38	4	50	other
m	s	20	3	73	180	20	20	360	republican
m	sr	21	3.4	72	170	25	1	120	other
m	j	20	3.2	72	150	6	14	240	republican
m	sr	23	3.2	72	180	16	8	240	other
m	s	23	3.1	75	175	21	2	150	republican
m	j	35	3	70	180	3	4	200	republican
m	sr	25	3.3	75	230	20	5	240	democrat
m	j	20	3.3	72	270	5	2	360	other
m	j	28	2.9	69	172	5	4	500	other
m	sr	26	3.4	75	200	4	1	45	other

gender	Class	Age	Gpa	Ht	Wt	Internet	phone	Exercise	softdrink
f	f	18	3.5	67	115	20	30	150	7up
f	f	18	3.9	66	130	10	3	90	drpepper
f	f	19	3.5	61	107	30	15	30	drpepper
f	f	18	3.6	70	142	13	16	680	other
f	f	24	3.6	67	150	8	1	300	other
f	g	29	3.3	66	155	3	3	420	coke
f	g	33	3.4	65	135	10	15	210	coke
f	g	34	3.7	66	110	5	3	600	other
f	g	34	3.7	66	149	7	2	480	pepsi
f	g	26	3	67	136	7	8	420	pepsi
f	g	54	3.9	76	107	1	2	135	pepsi
f	j	20	3.7	67	145	10	10	360	7up
f	j	20	3.1	64	118	7	6	180	coke
f	j	22	2.6	66	142	25	45	120	coke
f	j	22	2.7	62	130	25	2	0	drpepper
f	j	22	2.9	68	122	21	10	240	drpepper
f	j	21	3.9	64	130	7	1	45	drpepper
f	j	21	3.9	63	130	7	1	240	drpepper
f	j	20	3.6	64	110	30	2	0	other
f	j	21	3	67	115	10	10	0	other
f	j	20	3.9	67	150	10	7	280	other
f	j	27	3	67	180	5	2	60	other
f	j	20	2.9	63	170	16	3	150	other
f	j	19	3	67	150	7	2	180	other
f	j	21	3	62	145	10	3	420	pepsi
f	j	19	3.4	67	150	5	21	420	pepsi
f	s	19	3.6	60	97	24	14	30	7up
f	s	19	3.7	67	133	2	2	90	7up
f	s	19	3.2	68	130	7	2	300	7up
f	s	19	2.8	69	165	30	14	150	coke
f	s	19	4	63	112	4	3	300	drpepper
f	s	19	3.2	67	150	21	4	0	drpepper
f	s	20	3.2	67	130	3	1	70	other
f	s	18	3.3	68	155	10	2	60	other
f	s	23	2.3	63	106	9	2	120	other
f	s	25	3	63	115	40	3	120	other
f	sr	21	3.8	65	130	40	2	120	coke
f	sr	21	2.9	68	140	10	11	300	coke
f	sr	22	3.5	65	150	25	14	240	coke

f	sr	21	3.3	64	130	14	2	300	coke
f	sr	21	3	70	135	7	2	0	coke
f	sr	22	2.3	65	205	15	10	120	coke
f	sr	28	3.7	67	140	20	4	400	drpepper
f	sr	22	3.3	78	170	8	2	900	drpepper
f	sr	22	3.7	70	180	55	2	300	drpepper
f	sr	23	3.9	65	170	20	7	420	other
f	sr	22	3.9	66	135	14	2	180	other
f	sr	21	3.8	64	120	28	4	420	other
f	sr	24	3.6	64	127	15	2	360	pepsi
f	sr	30	2.6	62	131	20	7	60	pepsi
m	f	19	3.2	73	165	3	1	300	drpepper
m	g	24	3.3	72	190	4	2	90	drpepper
m	g	41	3.7	69	250	9	3	60	drpepper
m	j	21	3.3	72	205	15	20	420	7up
m	j	26	3.9	69	173	6	2	420	coke
m	j	21	3.4	72	170	3	3	90	coke
m	j	20	2.9	71	185	15	10	600	drpepper
m	j	23	1.5	72	135	6	2	180	drpepper
m	j	19	3.1	75	149	21	10	520	drpepper
m	j	21	2.8	71	185	20	24	600	drpepper
m	j	25	3	72	205	8	2	180	other
m	j	20	2.5	70	150	26	5	480	other
m	j	21	3.3	70	175	10	6	100	other
m	j	27	2.9	73	190	7	10	90	other
m	j	20	3.2	72	183	28	21	600	pepsi
m	j	21	2.4	74	180	7	1	225	pepsi
m	j	21	3.9	73	150	10	2	120	pepsi
m	s	19	3	74	184	10	3	360	coke
m	s	20	3.2	70	160	3	2	180	drpepper
m	s	20	3.8	66	150	2	3	240	other
m	s	43	3.8	74	270	7	3	300	pepsi
m	sr	22	3	72	180	5	2	300	coke
m	sr	22	3.9	71	185	3	1	270	drpepper
m	sr	22	3.9	73	225	10	1	300	drpepper
m	sr	22	2.7	73	148	12	10	420	drpepper
m	sr	43	2.7	76	210	5	1	90	other
m	sr	31	3.6	75	195	10	5	30	other
m	sr	22	3	75	180	40	8	200	other
m	sr	23	3.8	69	160	20	8	60	other
m	sr	21	2.8	70	155	3	7	120	pepsi

Summer12.xls

gender	class	age	gpa	Ht	wt	internet	phone	exercise	party
f	s	20	3.9	62	130	25	10	180	republican
f	sr	22	3	64	129	9	4	240	other
f	j	35	3.9	68	154	3	3	60	other
f	j	20	2.9	64	142	6	1	420	democrat
f	s	21	2.8	63	190	39	1	120	republican
f	j	20	3.1	67	150	12	3	360	other
f	j	20	3	67	170	20	10	560	other
f	g	25	3.6	68	250	24	50	90	republican
f	sr	21	3.9	66	160	10	10	120	democrat
f	j	24	3.5	65	163	5	1	240	democrat
f	f	19	3	67	160	9	4	30	other
f	f	18	3.2	67	150	10	7	240	other
f	g	48	3	63	200	10	7	270	democrat
f	j	21	3.8	66	165	4	2	300	republican
f	sr	26	2.6	65	150	25	2	60	democrat
f	f	36	3.9	68	200	7	4	160	other
f	f	18	4	62	110	10	5	120	other
f	f	18	3.9	68	135	10	25	650	republican
f	f	18	3.8	67	135	10	15	150	republican
f	f	19	3.7	64	115	7	14	120	republican
f	j	21	3.6	65	125	4	3	0	republican
f	sr	23	3.4	67	135	16	8	360	republican
f	f	18	3.5	74	145	5	2	240	other
f	f	18	3	68	120	7	8	120	republican
f	sr	27	2.9	65	135	2	2	60	democrat
f	s	20	3.8	61	96	6	3	60	other
f	j	25	3.2	67	135	10	20	600	democrat
f	s	31	2.5	62	125	10	1	0	democrat
f	f	18	3.5	65	133	20	20	700	republican
f	f	18	4	64	148	40	1	60	other
f	s	21	3.8	61	100	8	2	60	republican
f	s	20	2.6	67	160	28	15	0	other
f	g	44	4	67	168	4	2	120	other
f	s	20	3.7	63	145	8	1	180	other
f	s	20	3	63	150	18	2	120	other
f	sr	24	3.2	62	116	10	8	200	other
f	s	20	3.5	70	125	20	20	210	other
f	s	43	3.7	67	280	32	21	185	democrat
m	s	20	4	72	180	17	2	450	democrat

m	j	22	3.5	75	190	20	1	1200	other
m	j	20	3.1	64	120	10	1	60	republican
m	j	22	3.4	74	185	10	5	250	other
m	s	20	3	70	165	10	2	700	republican
m	j	22	3.2	71	109	38	4	50	other
m	s	20	3	73	180	20	20	360	republican
m	sr	21	3.4	72	170	25	1	120	other
m	j	20	3.2	72	150	6	14	240	republican
m	sr	23	3.2	72	180	16	8	240	other
m	s	23	3.1	75	175	21	2	150	republican
m	j	35	3	70	180	3	4	200	republican
m	sr	25	3.3	75	230	20	5	240	democrat
m	j	20	3.3	72	270	5	2	360	other
m	j	28	2.9	69	172	5	4	500	other
m	sr	26	3.4	75	200	4	1	45	other
f	s	20	2.9	63	140	20	2	500	other
m	sr	27	3.85	67	160	10	1	180	democrat
m	sr	30	3.6	72	180	6	1	300	other
m	s	19	3	72	130	14	30	600	republican
m	sr	32	2.9	76	300	8	1	120	republican
f	sr	20	2.9	69	139	10	15	240	democrat
m	sr	22	3.85	66	220	4	1	320	other
f	sr	24	4	64	140	10	7	380	republican
f	sr	41	3.3	64	240	32	24	140	other
f	s	24	2.7	66	150	16	3	180	other
f	f	19	3.4	64	115	20	15	300	republican
f	g	50	3.8	65	159	4	3	90	democrat
m	j	24	3.3	66	140	5	2	300	democrat
f	sr	21	3.3	61	120	30	30	60	republican
f	sr	21	3.75	67	160	25	10	100	other
f	g	27	3	67	160	20	10	30	other
f	sr	27	3.6	69	160	3	3	150	republican
m	sr	23	2.75	71	182	14	2	360	republican
m	sr	23	3.2	68	190	10	1	1500	democrat
m	sr	27	3.1	74	300	28	15	120	other

These files contain actual data collected on students in a statistical methods course. To obtain an electronic copy of these files send an e-mail to Dr. Larry Stephens(e-mail Lstephens@unomaha.edu)